Particles and Forces

Particles and Forces
At the Heart of Matter

. . .

READINGS FROM
SCIENTIFIC AMERICAN MAGAZINE

Edited by

Richard A. Carrigan, Jr.

Fermi National Accelerator Laboratory

and

W. Peter Trower

Virginia Polytechnic Institute and State University

W. H. FREEMAN AND COMPANY
New York

Some of the SCIENTIFIC AMERICAN articles in *Particles and Forces: At the Heart of Matter* are available as separate Offprints. For a complete list of articles now available as Offprints, write to Product Manager, Marketing Department, W. H. Freeman and Company, 41 Madison Avenue, New York, New York 10010.

Library of Congress Cataloging-in-Publication Data

Particles and forces: at the heart of matter: readings from
   Scientific American magazine / edited by Richard A.
   Carrigan, Jr. and W. Peter Trower.
     p.   cm.
   Includes bibliographical references.
   ISBN 0-7167-2070-1
   1. Particles (Nuclear physics) I. Carrigan, R. A.
II. Trower, W. Peter.   III. Scientific American.
QC793.2.P38   1990
593.7'2--dc20   89-17036
                                CIP

Printed in the United States of America

1 2 3 4 5 6 7 8 9 0 RRD 8 9

CONTENTS

SECTION III: WEAK INTERACTIONS

SECTION IV: STRONG INTERACTIONS

SECTION V: NOW AND BEYOND

Note on cross-references to SCIENTIFIC AMERICAN articles: Articles included in this book are referred to by chapter number and title; articles not included in this book but available as Offprints are referred to by title, date of publication and Offprint number; articles not in this book and not available as Offprints are referred to by title and date of publication.

Introduction

The past two decades have seen far-reaching breakthroughs in our understanding of fundamental physics. We can now explain much about the tiniest particles as well as the mighty universe itself. More amazing still, developments on staggeringly different scales now appear to be inexorably linked.

That protons and mesons are formed from fractionally charged quarks has gained universal acceptance. These quarks seem to be confined in particles by colored gluons. Two of the forces that govern quark behavior—electromagnetism and weak interactions or radioactivity—have been joined as one. So-called Grand Unification Theories offer hope for further unification with the strong nuclear forces. All of this understanding has resulted from research with powerful accelerators and increasingly sophisticated experimental detectors.

On the cosmic scale the discovery of the three-degree black-body radiation has led us to believe that the universe is remarkably uniform or homogeneous. Further, protons are rare compared with these three-degree photons that pervade the universe. The discovery in 1964 of the violation of previously sacrosanct time reversal symmetry (strictly CP violation) has provided a mechanism to explain this ratio and the absence of antimatter in the universe. However, in the late 1970's a fly was found in the ointment—the Grand Unification Theories (known as GUTs) predicted many heavy magnetic monopoles but few if any now seem to be there. A new inflationary cosmology has since solved the problem of too many monopoles and explained the large-scale uniformity of matter in the universe. Thus the early moments of the universe—the birth of matter, the world and the cosmos as we know them—appear to tightly link the biggest bang with the smallest particle.

During the past two decades SCIENTIFIC AMERICAN has skillfully presented these original developments, using the best possible reporters—the scientists generating them. The selection of SCIENTIFIC AMERICAN articles in this volume chronicles these latest developments in particle physics. Some of the authors are among the most distinguished physicists in the world today.

Particles and Forces: At the Heart of Matter explores the very smallest objects in the universe and their interactions. Chris Quigg begins with a bird's-eye view of the current state of particle physics theory. Sheldon Lee Glashow captures the early moments when the first unification of electromagnetism and the weak interactions occurred. Claudio Rebbi shows us how the important process of quark confinement may work in detail.

Key accelerator developments, the instruments that have made experimental investigations of many of these new principles possible, are summarized by Robert R. Wilson — one of the world's premier accelerator builders. The Superconducting Super Collider, or SSC, is an exciting prospect for the future about which J. David Jackson, Maury Tigner and Stanley Wojcicki, the leaders of the SSC design team, describe their vision for this stupendous accelerator.

For the weak interaction of particles, Martin L. Perl and William T. Kirk review the discovery of the tau, the newest of the heavy leptons, while David B. Cline, Carlo Rubbia and Simon van der Meer report on the development of an accelerator complex and the mammoth detectors that ultimately uncovered the $W^{\pm}$ and Z^0 mesons.

For strongly interacting particles Leon M. Lederman recounts the discovery of the upsilon — the heaviest of the "onium" particles and the first concrete evidence for a beauty quark. Elliott D. Bloom and Gary J. Feldman review quarkonium in detail, showing how it qualitatively follows the more familiar picture of the hydrogen atom. Nariman B. Mistry, Ronald A. Poling and Edward H. Thorndike follow with their recent work on naked beauty using the Cornell storage ring system.

What kinds of physics lie beyond the current standard model? Michael B. Green discusses the interesting developments in superstring theory that attempt to replace particles with very tiny strings interacting in a ten-dimensional space. In contrast, Haim Harari investigates the possibility of quarks and leptons arising out of some subquark system.

Particles and Forces: At the Heart of Matter is only half the story. Our other volume in this series, *Particle Physics in the Cosmos*, recounts an equally important part, the intimate interconnection between particle physics and the Big Bang. It is this link between the birth of the universe and the fundamental particles and forces described here that make these two subjects even more interesting. Indeed, the connection of these topics may be the most profound idea to emerge from the last half of the twentieth century.

Because the two subjects are so tightly linked, some important theoretical concepts are covered in more detail in *Particle Physics in the Cosmos*. In particular, the details of grand unification appear there.

Interested readers may want to have both volumes. Except for the introduction, there is no overlap.

A reprint volume is special by its nature. Each article in this volume was written independently, so there is some repetition. The repetition offers advantages. First, each author has a different point of view so that the reader can get an individual perspective from each article. The second advantage is that if one article seems too long or too abstract, the reader can try a different one. Sometimes it helps to read the experimental articles first for an easier view of theory. But be warned — some of the longest and most difficult articles also offer some of the greatest insights into the underlying physics. With the articles on the SSC, string theory and possible structure of quarks and leptons, we have tried to provide a glimpse over the horizon and into the future.

Another word of warning. Particle physicists have recently become enamored with "s" words, "standard" and "super" being the worst offenders. In this volume the standard model in one section may refer to electroweak unification, in another to some form of grand unification and in a third to cosmology.

For whom is this book intended? We have addressed four audiences: serious high-school students, university students, intelligent general readers and scientists themselves. While some of the articles are not simple, they offer the thoughtful reader a unique insight into the thinking of the scientists that did the work.

An ideal use for the book would be to give it to a superior high-school student and ask her to guess what will happen next. Our present picture of physics and cosmology is incomplete and some things are undoubtedly wrong. This is the fabric of science, but as Haim Harari says in the last chapter, "Imagining what a successful model might be like, however, is not at all the same thing as actually constructing a realistic and internally consistent one."

A significant feature of these articles is the guiding hands of the SCIENTIFIC AMERICAN editors and their skillful illustrators. The editor plays an important role in bringing the involved scientist down to traditional language. Most of the authors have felt that the editors drove them too far toward popularization, while their editors have believed the scientists could never appreciate the need for clarification. The result is a compromise.

To help with the interrelating and updating of these articles we have added postscripts, often by the authors themselves. These are minor emendations since most of the articles have stood the test of time remarkably well.

The editors of SCIENTIFIC AMERICAN have played an essential role by selecting and shaping the original articles. The articles were drawn from more than twice the number that appeared on the subject in SCIENTIFIC AMERICAN. Space limitations have forced us to omit some important articles. We have tried to balance experiment-theory, particle-cosmology, perspectives from labs around the world, our own prejudices and internal discord and other factors. To our friends and colleagues whose articles do not appear here, we can only say that the other of us, a most obdurate and unthinking person, was responsible.

Richard A. Carrigan, Jr.
W. Peter Trower

SECTION

IDEAS

. . .

Elementary Particles and Forces

A coherent view of the fundamental constituents of matter and the forces governing them has emerged. It embraces disparate theories, but they may soon be united in one comprehensive description of natural events.

. . .

Chris Quigg
April, 1985

The notion that a fundamental simplicity lies below the observed diversity of the universe has carried physics far. Historically the list of particles and forces considered to be elementary has changed continually as closer scrutiny of matter and its interactions revealed microcosms within microcosms: atoms within molecules, nuclei and electrons within atoms, and successively deeper levels of structure within the nucleus. Over the past decade, however, experimental results and the convergence of theoretical ideas have brought new coherence to the subject of particle physics, raising hopes that an enduring understanding of the laws of nature is within reach.

Higher accelerator energies have made it possible to collide particles with greater violence, revealing the subatomic realm in correspondingly finer detail; the limit of experimental resolution now stands at about 10^{-16} centimeter, about a thousandth the diameter of a proton. A decade ago physics recognized hundreds of apparently elementary particles; at today's resolution that diversity has been shown to represent combinations of a much smaller number of fundamental entities. Meanwhile the forces through which these constituents interact have begun to display underlying similarities. A deep connection between two of the forces, electromag-

netism and the weak force that is familiar in nuclear decay, has been established, and prospects are good for a description of fundamental forces that also encompasses the strong force that binds atomic nuclei.

Of the particles that now appear to be structureless and indivisible, and therefore fundamental, those that are not affected by the strong force are known as leptons. Six distinct types, fancifully called flavors, of lepton have been identified (see Figure 1.1). Three of the leptons, the electron, the muon and the tau, carry an identical electric charge of −1; they differ, however, in mass. The electron is the lightest and the tau the heaviest of the three. The other three, the neutrinos, are, as their name suggests, electrically neutral. Two of them, the electron neutrino and the muon neutrino, have been shown to be nearly massless. In spite of their varied masses all six leptons carry precisely the same amount of spin angular momentum. They are designated spin-1/2 because each particle can spin in one of two directions. A lepton is said to be right-handed if the curled fingers of a right hand indicate its rotation when the thumb points in its direction of travel and left-handed when the fingers and thumb of the left hand indicate its spin and direction.

For each lepton there is a corresponding antilepton, a variety of antiparticle. Antiparticles have the same mass and spin as their respective particles but carry opposite values for other properties, such as electric charge. The antileptons, for example, include the antielectron, or positron, the antimuon and the antitau, all of which are positively charged, and three electrically neutral antineutrinos.

In their interactions the leptons seem to observe boundaries that define three families, each composed of a charged lepton and its neutrino. The families are distinguished mathematically by lepton numbers; for example, the electron and the electron neutrino are assigned electron number 1, muon number 0 and tau number 0. Antileptons are assigned lepton numbers of the opposite sign. Although some of the leptons decay into other leptons, the total lepton number of the decay products is equal to that of the original particle; consequently the family lines are preserved.

The muon, for example, is unstable. It decays after a mean lifetime of 2.2 microseconds into an electron, an electron antineutrino and a muon neutrino through a process mediated by the weak force. Total lepton number is unaltered in the transformation. The muon number of the muon neutrino is 1, the electron number of the electron is 1 and that of the electron antineutrino is −1. The electron numbers cancel, leaving the initial muon number of 1 unchanged. Lepton number is also conserved in the decay of the tau, which endures for a mean lifetime of 3×10^{-13} second.

The electron, however, is absolutely stable. Electric charge must be conserved in all interactions, and there is no less massive charged particle into which an electron could decay. The decay of neutrinos has not been observed. Because neutrinos are the less massive members of their respective families, their decay would necessarily cross family lines.

Where are leptons observed? The electron is familiar as the carrier of electric charge in metals and semiconductors. Electron antineutrinos are emitted in the beta decay of neutrons into protons. Nuclear reactors, which produce large number of unstable free neutrons, are abundant sources of antineutrinos. The remaining species of lepton are produced mainly in high-energy collisions of subnuclear particles, which occur naturally as cosmic rays interact with the atmosphere and under controlled conditions in particle accelerators. Only the tau neutrino has not been observed directly, but the indirect evidence for its existence is convincing.

QUARKS

Subnuclear particles that experience the strong force make up the second great class of particles studied in the laboratory. These are the hadrons; among them are the protons, the neutrons and the mesons. A host of other less familiar hadrons exist only ephemerally as the products of high-energy collisions, from which extremely massive and very unstable particles can materialize. Hundreds of species of hadron have been catalogued, varying in mass spin, charge and other properties.

Hadrons are not elementary particles, however, since they have internal structure. In 1964 Murray Gell-Mann of the California Institute of Technology and George Zweig, then working at CERN, the European laboratory for particle physics in Geneva, independently attempted to account for the bewildering variety of hadrons by suggesting they are composite particles, each a different combination of a small number of fundamental constituents. Gell-Mann called them quarks. Studies at the Stanford Linear Accelerator Center (SLAC) in the late 1960's in which high-energy electrons were fired at protons and neutrons bolstered the hypothesis. The distribution in energy and angle of the scattered electrons indicated that some were colliding with pointlike, electrically charged objects within the protons and neutrons.

Particle physics now attributes all known hadron species to combinations of these fundamental entities. Five kinds, also termed flavors, of quark have been identified—the up (*u*), down (*d*), charm (*c*), strange (*s*) and bottom (*b*) quarks—and a sixth flavor, the top (*t*) quark, is believed to exist (see Figure 1.1). Like the leptons, quarks have half a unit of spin and can therefore exist in left-and right-handed states. They also carry electric charge equal to a precise fraction of an electron's charge: the *d*, *s* and *b* quarks have a charge of −1/3, and the *u*, *c* and the conjectured *t* quark have a charge of +2/3. The corresponding antiquarks have electric charges of the same magnitude but opposite sign.

Such fractional charges are never observed in hadrons, because quarks form combinations in which the sum of their charges is integral. Mesons, for example, consist of a quark and an antiquark, whose charges add up to −1, 0 or +1. Protons and neutrons consist respectively of two *u* quarks and a *d* quark, for a total charge of +1, and of a *u* quark and two *d* quarks, for a total charge of 0.

Like leptons, the quarks experience weak interac-

LEPTONS				QUARKS			
PARTICLE NAME	SYMBOL	MASS AT REST (MeV)	ELECTRIC CHARGE	PARTICLE NAME	SYMBOL	MASS AT REST (MeV)	ELECTRIC CHARGE
ELECTRON NEUTRINO	ν_e	ABOUT 0	0	UP	u	310	$2/3$
ELECTRON	e or e^-	0.511	-1	DOWN	d	310	$-1/3$
MUON NEUTRINO	ν_μ	ABOUT 0	0	CHARM	c	1,500	$2/3$
MUON	μ or μ^-	106.6	-1	STRANGE	s	505	$-1/3$
TAU NEUTRINO	ν_τ	LESS THAN 164	0	TOP/TRUTH	t	> 22,500; HYPOTHETICAL PARTICLE	$2/3$
TAU	τ or τ^-	1,784	-1	BOTTOM/BEAUTY	b	ABOUT 5,000	$-1/3$

FORCE	RANGE	STRENGTH AT 10^{-13} CENTIMETER IN COMPARISON WITH STRONG FORCE	CARRIER	MASS AT REST (GeV)	SPIN	ELECTRIC CHARGE	REMARKS
GRAVITY	INFINITE	10^{-38}	GRAVITON	0	2	0	CONJECTURED
ELECTROMAGNETISM	INFINITE	10^{-2}	PHOTON	0	1	0	OBSERVED DIRECTLY
WEAK	LESS THAN 10^{-16} CENTIMETER	10^{-13}	INTERMEDIATE BOSONS: W^+	81	1	$+1$	OBSERVED DIRECTLY
			W^-	81	1	-1	OBSERVED DIRECTLY
			Z^0	93	1	0	OBSERVED DIRECTLY
STRONG	LESS THAN 10^{-13} CENTIMETER	1	GLUONS	0	1	0	PERMANENTLY CONFINED

Figure 1.1 FUNDAMENTAL SCHEME OF NATURE embraces 12 elementary particles (*top*) and four forces (*bottom*). Among the properties of the particles are an identical amount of spin (1/2) and differing values of electric charge, color charge and mass (given as energy in millions of electron volts or MeV). Only the pairs of leptons and quarks at the top of each column are found in ordinary matter. The four forces thought to govern matter vary in range and strength. All the forces are conveyed by force particles, whose masses are given in billions of electron volts (GeV).

tions that change one species, or flavor, into another. For example, in the beta decay of a neutron into a proton one of the neutron's *d* quarks metamorphoses into a *u* quark, emitting an electron and an antineutrino in the process. Similar transformations of *c* quarks into *s* quarks have been observed. The pattern of decays suggests two family groupings, one of them thought to contain the *u* and the *d* quarks and the second the *c* and the *s* quarks. In apparent contrast to the behavior of leptons, some quark decays do cross family lines, however; transformations of *u* quarks into *s* quarks and of *c* quarks into *d* quarks have been observed. It is the similarity of the two known quark families to the families of leptons that first suggested the existence of a *t* quark, to serve as the partner of the *b* quark in a third family.

In contrast to the leptons, free quarks have never been observed. Yet circumstantial evidence for their existence has mounted steadily. One indication of the soundness of the quark model is its success in predicting the outcome of high-energy collisions of an electron and a positron. Because they represent matter and antimatter, the two particles annihilate each other, releasing energy in the form of a photon. The quark model predicts that the energy of the photon can materialize into a quark and an antiquark. Because the colliding electron-positron pair had a net momentum of 0, the quark-antiquark pair must diverge in opposite directions at equal velocities so that their net momentum is also 0. The quarks themselves go unobserved because their energy is converted into additional quarks and antiquarks, which materialize and combine with the original pair, giving rise to two jets of hadrons (most of them pions, a species of meson). Such jets are indeed observed, and their focused nature confirms that the hadrons did not arise directly from the

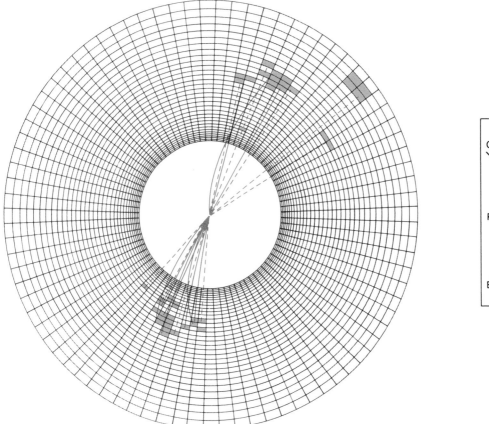

Figure 1.2 TWO NARROW JETS OF PARTICLES emerge from the collision and mutual annihilation of an electron and a positron. The annihilation releases energy, which gives rise to matter. Some of the detected particles are neutral (*broken lines*) and some electrically charged (*solid lines*). The focused character of the jets suggests that each jet developed from a single precursor: a quark or an anti-quark. They are the immediate products of the photon of electromagnetic energy released in the collision, which is diagrammed at the right. The paths of the particles were reconstructed by computer from ionization tracks and from the pattern of energy (*color*) deposited as the particles struck the inner layer of the detector.

collision but from single, indivisible particles whose trajectories the jets preserve. Figure 1.2 shows such an event recorded in the JADE detector of the PETRA accelerator at the Deutsches Elektronen-Synchroton (DESY) in Hamburg.

The case for the reality of quarks is also supported by the variety of energy levels, or masses, at which certain species of hadron, notably the psi and the upsilon particles, can be observed in accelerator experiments (see Figure 1.3). Such energy spectra appear analogous to atomic spectra: they seem to represent the quantum states of a bound system of two smaller components. Each of its quantum states would represent a different degree of excitation and a different combination of the components' spins and orbital motion. To most physicists the conclusion that such particles are made up of quarks is irresistible. The psi particle is held to consist of a *c* quark and its antiquark, and the upsilon particle is believed to comprise a *b* quark and its antiquark.

What rules govern the combinations of quarks that form hadrons? Mesons are composed of a quark and an antiquark. Because each quark has a spin of 1/2, the net spin of a meson is 0 if its constituents spin in opposite directions and 1 if they spin in the same direction, although in their excited

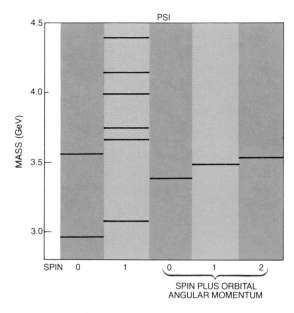

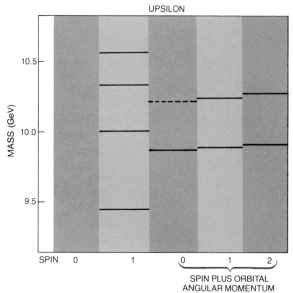

Figure 1.3 VARIETY OF MASSES at which two-quark systems known as the psi (*left*) and the upsilon (*right*) particles are observed reveal the energy states each can adopt. The psi particle consists of a *c* quark and its antiquark, bound by the color force; the upsilon particle is a similar combination of a *b* quark and its antiquark. Each column within a spectrum of masses corresponds to a different combination of the quarks' spins and orbital angular momentum. Different masses or, equivalently, energy levels within a column represent quantum levels of excitation. The resemblance of the spectra to the spectra of atoms indicates that particles such as the psi and the upsilon are also bound systems of smaller constituents. Such spectra offer insight into the behavior of the color force at short distances.

states mesons may have larger values of spin owing to the quarks' orbital motion. The other class of hadrons, the baryons, consist of three quarks each. Summing the constituent quarks' possible spins and directions yields two possible values for the spin of the least energetic baryons: 1/2 and 3/2. No other combinations of quarks have been observed; hadrons that consist of two or four quarks seem to be ruled out.

The reason is linked with the answer to another puzzle. According to the exclusion principle of Wolfgang Pauli, no two particles occupying a minute region of space and possessing half-integral spins can have the same quantum number—the same values of momentum, charge and spin. The Pauli exclusion principle accounts elegantly for the configurations of electrons that determine an element's place in the periodic table. We should expect it to be a reliable guide to the panoply of hadrons as well. The principle would seem to suggest, however, that exotic hadrons such as the delta plus plus and the omega minus particles, which materialize briefly following high-energy collisions, cannot exist. They consist respectively of three *u* and three *s* quarks and possess a spin of 3/2; all three quarks in each of the hadrons must be identical in spin as well as in other properties and hence must occupy the same quantum state.

COLORS

To explain such observed combinations it is necessary to suppose the three otherwise identical quarks are distinguished by another trait: a new kind of charge, whimsically termed color, on which the strong force acts. Each flavor of quark can carry one of three kinds of color charge: red, green or blue. To a red quark there corresponds an antiquark with a color charge of antired (which may be thought of as cyan); other antiquarks bear charges of antigreen (magenta) and antiblue (yellow).

The analogy between this new kind of charge and color makes it possible to specify the rules under which quarks combine. Hadrons do not exhibit a color charge; the sum of the component quarks' colors must be white, or color-neutral. Therefore the only allowable combinations are those of a quark and its antiquark, giving rise to mesons, and of a red, a green and a blue quark, yielding the baryons.

Colored states are never seen in isolation. This concealment is consistent with the fact that free quarks, bearing a single color charge, have never been observed. The activity of the strong force between colored quarks must be extraordinarily powerful, perhaps powerful enough to confine quarks permanently within colorless, or color-neutral, hadrons. The description of violent electron-positron collisions according to the quark model, however, assumes the quarks that give rise to the observed jets of hadrons diverge freely during the first instant following the collision. The apparent independence of quarks at very short distances is known as asymptotic freedom; it was described in 1973 by David J. Gross and Frank Wilczek of Princeton University and by H. David Politzer, then at Harvard University.

Analogy yields an operational understanding of this paradoxical state of affairs, in which quarks interact only weakly when they are close together and yet cannot be separated. We may think of a hadron as a bubble within which quarks are imprisoned. Within the bubble the quarks move freely, but they cannot escape from it. The bubbles, of course, are only a metaphor for the dynamical behavior of the force between quarks, and a fuller explanation for what is known as quark confinement can come only from an examination of the forces through which particles interact.

THE FUNDAMENTAL INTERACTIONS

Nature contrives enormous complexity of structure and dynamics from the six leptons and six quarks now thought to be the fundamental constituents of matter. Four forces govern their relations: electromagnetism, gravity and the strong and weak forces. In the larger world we experience directly, a force can be defined as an agent that alters the velocity of a body by changing its speed or direction. In the realm of elementary particles, where quantum mechanics and relativity replace the Newtonian mechanics of the larger world, a more comprehensive notion of force is in order, and with it a more general term, interaction. An interaction can cause changes of energy, momentum or kind to occur among several colliding particles; an interaction can

also affect a particle in isolation, in a spontaneous decay process.

Only gravity has not been studied at the scale on which elementary particles exist; its effects on such minute masses are so small that they can safely be ignored. Physicists have attempted with considerable success to predict the behavior of the other three interactions through mathematical descriptions known as gauge theories.

The notion of symmetry is central to gauge theories. A symmetry, in the mathematical sense, arises when the solutions to a set of equations remain the same even though a characteristic of the system they describe is altered. If a mathematical theory remains valid when a characteristic of the system is changed by an identical amount at every point in space, it can be said that the equations display a global symmetry with respect to that characteristic. If the characteristic can be altered independently at every point in space and the theory is still valid, its equations display local symmetry with respect to the characteristic.

Each of the four fundamental forces is now thought to arise from the invariance of a law of nature, such as the conservation of charge or energy, under a local symmetry operation, in which a certain parameter is altered independently at every point in space. An analogy with an ideal rubber disk may help to visualize the effect of the mathematics. If the shape of the rubber disk is likened to a natural principle and the displacement of a point within the disk is regarded as a local symmetry operation, the disk must keep its shape even as each point within it is displaced independently. The displacements stretch the disk and introduce forces between points. Similarly, in gauge theories the fundamental forces are the inevitable consequences of local symmetry operations; they are required in order to preserve symmetry.

Of the three interactions studied in the realm of elementary particles, only electromagnetism is the stuff of everyday experience, familiar in the form of sunlight, the spark of a static discharge and the gentle swing of a compass needle. On the subatomic level it takes on an unfamiliar aspect. Accordingly to relativistic quantum theory, which links matter and energy, electromagnetic interactions are mediated by photons: massless "force particles" that embody precise quantities of energy. The quantum theory of electromagnetism, which describes the photon-mediated interactions of electrically charged particles, is known as quantum electrodynamics (QED).

In common with other theories of the fundamental interactions, QED is a gauge theory. In QED the electromagnetic force can be derived by requiring that the equations describing the motion of a charged particle remain unchanged in the course of local symmetry operations. Specifically, if the phase of the wave function by which a charged particle is described in quantum theory is altered independently at every point in space, QED requires that the electromagnetic interaction and its mediating particle, the photon, exist in order to maintain symmetry.

QED is the most successful of physical theories. Using calculation methods developed in the 1940's by Richard P. Feynman and others, it has achieved predictions of enormous accuracy, such as the infinitesimal effect of the photons radiated and absorbed by an electron on the magnetic moment generated by the electron's innate spin. Moreover, QED's descriptions of the electromagnetic interaction have been verified over an extraordinary range of distances, varying from less than 10^{-18} meter to more than 10^8 meters.

SCREENING

In particular QED has explained the effective weakening of the electromagnetic charge with distance. The electric charge carried by an object is a fixed and definite quantity. When a charge is surrounded by other freely moving charges, however, its effects may be modified. If an electron enters a medium composed of molecules that have positively and negatively charged ends, for example, it will polarize the molecules. The electron will repel their negative ends and attract their positive ends, in effect screening itself in positive charge (see Figure 1.4). The result of the polarization is to reduce the electron's effective charge by an amount that increases with distance. Only when the electron is inspected at very close range — on a submolecular scale, within the screen of positive charges — is its full charge apparent.

Such a screening effect seemingly should not arise in a vacuum, in which there are no molecules to become polarized. The uncertainty principle of Werner Heisenberg suggests, however, that the vacuum is not empty. According to the principle, uncertainty about the energy of a system increases as it is examined on progressively shorter time scales. Particles may violate the law of the conservation of energy for unobservably brief instants; in effect, they may materialize from nothingness. In QED the vacuum is seen as a complicated and seething medium in which pairs of charged "virtual" particles,

particularly electrons and positrons, have a fleeting existence. These ephemeral vacuum fluctuations are polarizable just as are the molecules of a gas or a liquid. Accordingly QED predicts that in a vacuum too electric charge will be screened and effectively reduced at large distances.

The strong interaction affecting quarks that is based on the color charge also varies with distance, although in a contrary manner: instead of weakening with distance the color charge appears to grow stronger. Only at distances of less than about 10^{-13} centimeter, the diameter of a proton, does it diminish enough to allow mutually bound quarks a degree of independence. Yet the explanation for this peculiar behavior is found in a theory that is closely modeled on QED. It is a theory called quantum chromodynamics (QCD), the gauge theory of the strong interactions.

Like QED, QCD postulates force particles, which mediate interactions. Colored quarks interact through the exchange of entities called gluons, just as charged particles trade photons. Whereas QED recognizes only one kind of photon, however, QCD admits eight kinds of gluon. In contrast to the photons of QED, which do not alter the charge of interacting particles, the emission or absorption of a gluon can change a quark's color; each of the eight gluons mediates a different transformation. The mediating gluon is itself colored, bearing both a color and an anticolor.

The fact that the gluons are color-charged, in contrast to the electrically neutral photons of QED, accounts for the differing behaviors over distance of the electromagnetic and strong interactions. In QCD two competing effects govern the effective charge: screening, analogous to the screening of QED, and a new effect known as camouflage. The screening, or vacuum polarization, resembles that in electromagnetic interactions. The vacuum of QCD is populated by pairs of virtual quarks and antiquarks, winking into and out of existence. If a quark is introduced into the vacuum, virtual particles bearing contrasting color charges will be attracted to the quark; those bearing a like charge will be repelled. Hence the quark's color charge will be hidden within a cloud of unlike colors, which serves to reduce the effective charge of the quark at greater distances.

CAMOUFLAGE

Within this polarized vacuum, however, the quark itself continuously emits and reabsorbs gluons, thereby changing its color. The color-charged gluons propagate to appreciable distances. In effect they spread the color charge throughout space, thus camouflaging the quark that is the source of the charge (see Figure 1.4). The smaller an arbitrary region of space centered on the quark is, the smaller will be the proportion of the quark's charge contained in it. Thus the color charge felt by a quark of another color will diminish as it approaches the first quark. Only at a large distance will the full magnitude of the color charge be apparent.

In QCD the behavior of the strong force represents the net effect of screening and camouflage. The equations of QCD yield a behavior that is consistent with the observed paradox of quarks: they are both permanently confined and asymptotically free. The strong interaction is calculated to become extraordinarily strong at appreciable distances, resulting in quark confinement, but to weaken and free quarks at very close range.

In the regime of short distances that is probed in high-energy collisions, strong interactions are so enfeebled that they can be described using the methods developed in the context of QED for the much weaker electromagnetic interaction. Hence some of the same precision that characterizes QED can be imparted to QCD. The evolution of jets of

hadrons from a quark and an antiquark generated in electron-positron annihilation, for example, is a strong interaction. QCD predicts that if the energy of the collision is high enough, the quark and the antiquark moving off in opposite directions may generate not two but three jets of hadrons. One of the particles will radiate a gluon, moving in a third direction. It will also evolve into hadrons, giving rise to a third distinct jet — a feature that indeed is commonly seen in high-energy collisions. (See Figure 1.5 recorded in the DESY JADE detector.)

The three jets continue along paths set by quarks and gluons moving within an extremely confined space, less than 10^{-13} centimeter. The quark-antiquark pair cannot proceed as isolated particles beyond that distance, the limit of asymptotic freedom. Yet the confinement of quarks and of their interactions is not quite absolute. Although a hadron as a whole is color-neutral, its quarks do respond to the individual color charges of quarks in neighboring hadrons. The interaction, feeble compared with the color forces within hadrons, generates the binding force that holds the protons together in nuclei.

Moreover, it seems likely that when hadronic matter is compressed and heated to extreme temper-

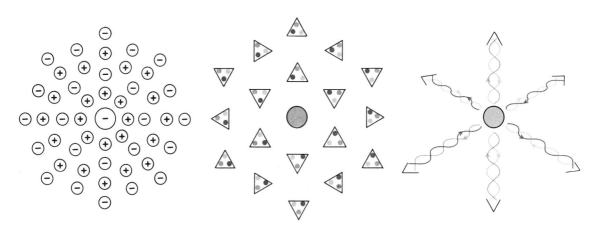

Figure 1.4 SCREENING AND CAMOUFLAGE. An electron in a vacuum (*left*) is surrounded by short-lived pairs of virtual electrons and positrons, which in quantum theory populate the vacuum. The electron attracts the virtual positrons and repels the virtual electrons, thereby screening itself. The farther from the electron a real charge is, the thicker the intervening screen and the smaller the electron's effective charge. The color force is subject to the same screening effect (*center*). Virtual color charges fill the vacuum; a colored quark attracts contrasting colors, thereby surrounding itself with a screen that reduces its effective charge with distance. Camouflage counteracts screening, however. A quark continuously radiates and reabsorbs gluons that carry its color charge and change its color, in this case from blue to green (*right*). Camouflage acts to increase the force felt by an actual quark as it moves toward the edge of the color-charged region.

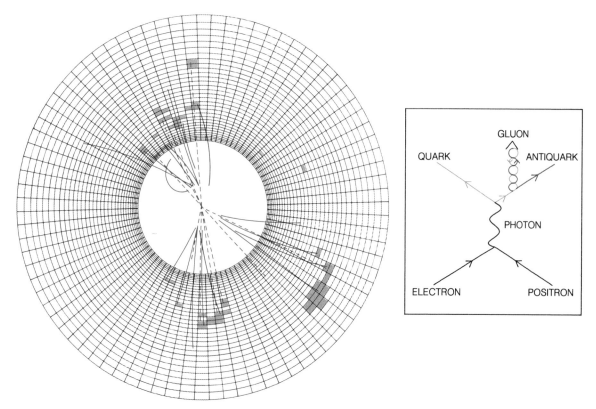

Figure 1.5 THREE-JET EVENT confirms the existence of the gluon, the mediating particle of the color force. An electron and a positron collided at high energy, creating a quark and an antiquark as in the event shown in Figure 1.2. In this case one of the quarks radiated a gluon (*right*). The quarks and the gluon diverged; each promptly gave rise to a shower of particles, which preserved the trajectory of the original entity (*left*). The event reveals the asymptotic freedom of quarks and gluons: their ability to move independently within a very small region in spite of the enormous strength of the color force across larger distances.

atures, the hadrons lose their individual identities. The hadronic bubbles of the image used above overlap and merge, possibly freeing their constituent quarks and gluons to migrate over great distances. The resulting state of matter, called quark-gluon plasma, may exist in the cores of collapsing supernovas and in neutron stars. Workers are now studying the possibility of creating quark-gluon plasma in the laboratory through collisions of heavy nuclei at very high energy [see "Hot Nuclear Matter," by Walter Greiner and Horst Stöcker; SCIENTIFIC AMERICAN, January, 1985].

ELECTROWEAK SYMMETRY

Understanding of the third interaction that elementary-particle physics must reckon with, the weak interaction, also has advanced by analogy with QED. In 1933 Enrico Fermi constructed the first mathematical description of the weak interaction, as manifested in beta radioactivity, by direct analogy with QED. Subsequent work revealed several important differences between the weak and the electromagnetic interactions. The weak force acts only over distances of less than 10^{-16} centimeter (in contrast to the long range of electromagnetism), and it is intimately associated with the spin of the interacting particles. Only particles with a left-handed spin are affected by weak interactions in which electric charge is changed, as in the beta decay of a neutron, whereas right-handed ones are unaffected.

In spite of these distinctions theorists extended the analogy and proposed that the weak interaction, like electromagnetism, is carried by a force particle, which came to be known as the intermediate boson, also called the *W* (for weak) particle. In order to mediate decays in which charge is changed, the *W* boson would need to carry electric charge. The range of a force is inversely proportional to the mass of the particle that transmits it; because the photon is massless, the electromagnetic interaction can act over infinite distances. The very short range of the weak force suggests an extremely massive boson.

A number of apparent connections between electromagnetism and the weak interaction, including the fact that the mediating particle of weak interactions is electrically charged, encouraged some workers to propose a synthesis. One immediate result of the proposal that the two interactions are only different manifestations of a single underlying phenomenon was an estimate for the mass of the *W* boson. The proposed unification implied that at very short distances and therefore at very high energies the weak force is equal to the electromagnetic force. Its apparent weakness in experiments done at lower energies merely reflects its short range. Therefore the whole of the difference in the apparent strengths of the two interactions must be due to the mass of the *W* boson. Under that assumption the *W* boson's mass can be estimated at about 100 times the mass of the proton.

To advance from the notion of a synthesis to a viable theory unifying the weak and the electromagnetic interactions has required half a century of experiments and theoretical insight, culminating in the work for which Sheldon Lee Glashow and Steven Weinberg, then at Harvard University, and Abdus Salam of the Imperial College of Science and Technology in London and the International Center for Theoretical Physics in Trieste won the 1979 Nobel prize in physics. Like QED itself, the unified, or electroweak, theory is a gauge theory derived from a symmetry principle, one that is manifested in the family groupings of quarks and leptons.

Not one but three intermediate bosons, along with the photon, serve as force particles in electroweak theory. They are the positively charged W^+ and negatively charged W^- bosons, which respectively mediate the exchange of positive and negative charge in weak interactions, and the Z^0 particle, which mediates a class of weak interactions known as neutral current processes. Neutral current processes such as the elastic scattering of a neutrino from a proton, a weak interaction in which no charge is exchanged, were predicted by the electroweak theory and first observed at CERN in 1973.

They represent a further point of convergence between electromagnetism and the weak interaction in that electromagnetic interactions do not change the charge of participating particles either.

To account for the fact that the electromagnetic and weak interactions, although they are intimately related, take different guises, the electroweak theory holds that the symmetry uniting them is apparent only at high energies. At lower energies it is concealed. An analogy can be drawn to the magnetic behavior of iron. When iron is warm, its molecules, which can be regarded as a set of infinitesimal magnets, are in hectic thermal motion and therefore randomly oriented. Viewed in the large the magnetic behavior of the iron is the same from all directions, reflecting the rotational symmetry of the laws of electromagnetism. When the iron cools below a critical temperature, however, its molecules line up in an arbitrary direction, leaving the metal magnetized along one axis. The symmetry of the underlying laws is now concealed.

The principal actor in the breaking of the symmetry that unites electromagnetism and the weak interaction at high energies is a postulated particle called the Higgs boson. It is through interactions with the Higgs boson that the symmetry-hiding masses of the intermediate bosons are generated. The Higgs boson is also held to be responsible for the fact that the quarks and leptons within the same family have different masses. At very high energies all quarks and leptons are thought to be massless; at lower energies interactions with the Higgs particle confer on the quarks and leptons their varying masses. Because the Higgs boson is elusive and may be far more massive than the intermediate bosons themselves, experimental energies much higher than those of current accelerators probably will be needed to produce it.

The three intermediate bosons required by the electroweak theory, however, have been observed. Energies high enough to produce such massive particles are best obtained in head-on collisions of protons and antiprotons. In one out of about five million collisions a quark from the proton and an antiquark from the antiproton fuse, yielding an intermediate boson. The boson disintegrates less than 10^{-24} second after its formation. Its brief existence, however, can be detected from its decay products.

In a triumph of accelerator art, experimental technique and theoretical reasoning, international teams at the CERN Super Proton Synchrotron Collider led by Carlo Rubbia of Harvard and Pierre Darriulat

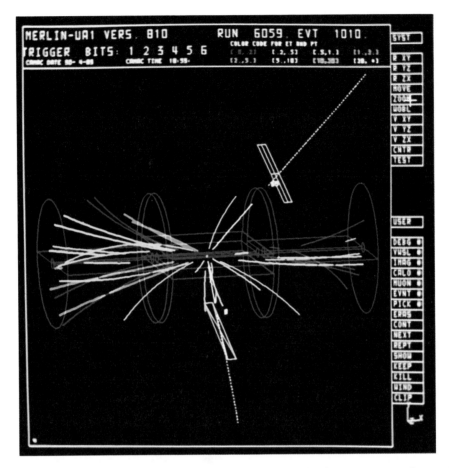

Figure 1.6 SIGNATURE OF THE Z^0 PARTICLE is visible on a computer-synthesized display in the CERN UA1 detector. The Z^0, W^+ and W^- are the particles that transmit the weak force; their existence was predicted by the unified theory of the weak and the electromagnetic interactions, and their discovery vindicated the theory. The tracks depicted within the detector correspond to particles detected following the high-energy collision of a proton and an antiproton. The tracks displayed in white are those of an electron and a positron, the characteristic decay products of the Z^0, which disintegrated soon after it materialized in the collision. (Brian Wolfe Photography.)

devised experiments that in 1983 detected the W bosons and the Z^0 particle (see Figure 1.6). An elaborate detector identified and recorded in the debris of violent proton-antiproton collisions single electrons whose trajectories matched the ones expected in W^- particle decays; the detector also recorded electrons and positrons traveling in precisely opposite directions, unmistakable evidence of the Z^0 particle. For their part in the experiments and in the design and construction of the proton-antiproton collider and the detector Rubbia and Simon van der Meer of CERN were awarded the 1984 Nobel prize in physics.

UNIFICATION

With QCD and the electroweak theory in hand, what remains to be understood? If both theories are correct, can they also be complete? Many observations are explained only in part, if at all, by the separate theories of the strong and the electroweak interactions. Some of them seem to invite a further unification of the strong, weak and electromagnetic interactions.

Among the hints of deeper patterns is the striking resemblance of quarks and leptons. Particles in both groups are structureless at current experimental resolution. Quarks possess color charges whereas leptons do not, but both carry a half unit of spin and take part in electromagnetic and weak interactions. Moreover, the electroweak theory itself suggests a relation between quarks and leptons. Unless each of the three lepton families (the electron and its neutrino, for example) can be linked with the corresponding family of quarks (the u and d quarks, in their three colors) the electroweak theory will be beset with mathematical inconsistencies.

What is known about the fundamental forces also points to a unification. All three can be described by gauge theories, which are similar in their mathematical structure. Moreover, the strengths of the three forces appear likely to converge at very short distances, a phenomenon that would be apparent only at extremely large energies. We have seen that the electromagnetic charge grows strong at short distances, whereas the strong, or color, charge becomes increasingly feeble. Might all the interactions become comparable at some gigantic energy?

If the interactions are fundamentally the same, the distinction between quarks, which respond to the strong force, and leptons, which do not, begins to dissolve. In the simplest example of a unified theory, put forward by Glashow and Howard Georgi of Harvard in 1974, each matched set of quarks and leptons gives rise to an extended family containing all the various states of charge and spin of each of the particles.

The mathematical consistency of the proposed organization of matter is impressive. Moreover, regularities in the scheme require that electric charge be apportioned among elementary particles in multiples of exactly 1/3, thereby accounting for the elec-

trical neutrality of stable matter. The atom is neutral only because when quarks are grouped in threes, as they are in the nucleus, their individual charges combine to give a charge that is a precise integer, equal and opposite to the charge of an integral number of electrons. If quarks were unrelated to leptons, the precise relation of their electric charges could only be a remarkable coincidence.

In such a unification only one gauge theory is required to describe all the interactions of matter. In a gauge theory each particle in a set can be transformed into any other particle. Transformations of quarks into other quarks and of leptons into other leptons, mediated by gluons and intermediate bosons, are familiar. A unified theory suggests that quarks can change into leptons and vice versa. As in any gauge theory, such an interaction would be mediated by a force particle: a postulated X or Y boson. Figure 1.7 illustrates this for one branch of a unified family of elementary particles. Like other gauge theories, the unified theory describes the variation over distance of interaction strengths. According to the simplest of the unified theories, the separate strong and electroweak interactions converge and become a single interaction at a distance of 10^{-29} centimeter, corresponding to an energy of 10^{24} electron volts.

Such an energy is far higher than may ever be attained in an accelerator, but certain consequences of unification might be apparent even in the low-energy world we inhabit. The supposition that transformations can cross the boundary between quarks and leptons implies that matter, much of whose mass consists of quarks, can decay. If, for example, the two u quarks in a proton were to ap-

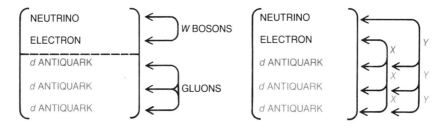

Figure 1.7 KINSHIP OF ALL MATTER is implied by unified theories of the fundamental forces. Because leptons respond to the electroweak force alone and quarks also respond to the strong force, the two are not equivalent in current theory, and transformations of one into the other have not been observed (*left*). If the simplest unified theories are correct and the fundamental forces are ultimately identical, then at some very high energy quarks and leptons are interconvertible (*right*). Known transformations are mediated by force particles such as the W bosons and gluons, transitions between quark and lepton groups by new force particles (X and Y).

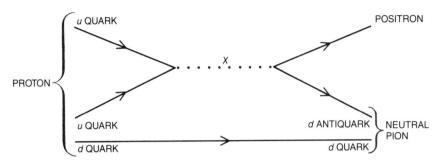

Figure 1.8 DECAY OF THE PROTON is a possible consequence of the transformations of quarks into leptons, a phenomenon unified theories would allow. The diagram shows one of several proposed decay routes. The proton's constituent *u* quarks combine to form an X particle, which disintegrates into a *d* antiquark and a positron (a lepton). The *d* antiquark combines with the remaining quark of the proton, a *d* quark, to form a neutral pion. Because pions are composed of matter and antimatter, they are short-lived; the mutual annihilation of their constituents will release energy in the form of two photons. The positron is also ephemeral: an encounter with a stray electron, its antiparticle, will convert it into energy as well.

proach each other closer than 10^{-29} centimeter, they might combine to form an X boson, which would disintegrate into a positron and a *d* antiquark. The antiquark would then combine with the one remaining quark of the proton, a *d* quark, to form a neutral pion, which itself would quickly decay into two photons. In the course of the process much of the proton's mass would be converted into energy (see Figure 1.8).

The observation of proton decay would lend a considerable support to a unified theory. It would also have interesting cosmological consequences. The universe contains far more matter than it does antimatter. Since matter and antimatter are equivalent in almost every respect, it is appealing to speculate that the universe was formed with equal amounts of both. If the number of baryons—three-quark particles such as the proton and the neutron, which constitute the bulk of ordinary matter—can change, as the decay of the proton would imply, then the current excess of matter need not represent the initial state of the universe. Originally matter and antimatter may indeed have been present in equal quantities, but during the first instants after the big bang, while the universe remained in a state of extremely high energy, processes that alter baryon number may have upset the balance.

A number of experiments have been mounted to search for proton decay. The large unification energy implies that the mean lifetime of the proton must be extraordinarily long—10^{30} years or more. To have a reasonable chance of observing a single

decay it is necessary to monitor an extremely large number of protons; a key feature of proton-decay experiments has therefore been large scale. The most ambitious experiment mounted to date is an instrumented tank of purified water 21 meters on a side in the Morton salt mine near Cleveland. During almost three years of monitoring none of the water's more than 10^{33} protons has been observed to decay, suggesting that the proton's lifetime is even longer than the simplest unified theory predicts. In some rival theories, however, the lifetime of the proton is considerably longer, and there are other theories in which protons decay in ways that would be difficult to detect in existing experiments. Furthermore, results from other experiments hint that protons can indeed decay.

OPEN QUESTIONS

Besides pointing the way to a possible unification the standard model, consisting of QCD and the electroweak theory, has suggested numerous sharp questions for present and future accelerators. Among the many goals for current facilities is an effort to test the predictions of QCD in greater detail. Over the next decade accelerators with the higher energies needed to produce the massive W and Z^0 bosons in adequate numbers will also add detail to electroweak theory. It would be presumptuous to say these investigations will turn up no surprises. The consistency and experimental successes of the standard model at familiar energies strongly suggest, however, that to resolve fundamental issues we need to take a large step up in interaction energy from the several hundred GeV (billion electron volts) attainable in the most powerful accelerators now being built.

Although the standard model is remarkably free of inconsistencies, it is incomplete; one is left hungry for further explanation. The model does not account for the pattern of quark and lepton masses or for the fact that although weak transitions usually observe family lines they occasionally cross them. The family pattern itself remains to be explained. Why should there be three matched sets of quarks and leptons? Might there be more?

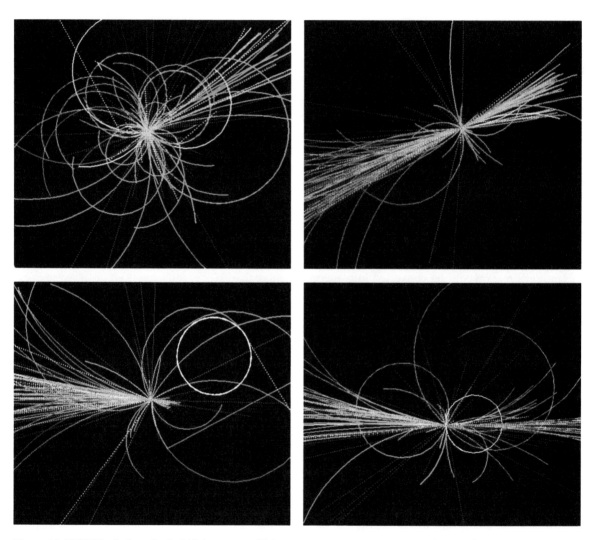

Figure 1.9 DEBRIS of a hypothetical high-energy collision between two protons is depicted in computer simulations made in accordance with the known and conjectured behavior of elementary particles. The collision takes place at an energy of 40 TeV (trillion electron volts), far greater than can be reached in today's accelerators. The enormous energy is assumed to give rise to a Higgs boson, which has not been observed. The Higgs boson promptly decays into two W bosons, also short-lived and massive, which then decay by several routes. Some of the particles whose tracks are plotted are the products of the W bosons' decay; others emerged from the breakup of the incident protons. A magnetic field is simulated, causing paths of charged particles to curve, whereas neutral ones are not affected. (Simulations by J. Freeman of Fermilab.)

Twenty or more parameters, constants not accounted for by theory, are required to specify the standard model completely. These include the coupling strengths of the strong, weak and electromagnetic interactions, the masses of the quarks and leptons, and parameters specifying the interactions of the Higgs boson. Furthermore, the apparently fundamental constituents and force particles number at least 34: 15 quarks (five flavors, each in three colors), six leptons, the photon, eight gluons, three intermediate bosons and the postulated Higgs boson. By the criterion of simplicity the standard model does not seem to represent progress over the ancient view of matter as made up of earth, air, fire and water, interacting through love and strife. Encouraged by historical precedent, many physicists account for the diversity by proposing that these seemingly fundamental particles are made up of still smaller particles in varying combinations.

There are two other crucial points at which the standard model seems to falter. Neither the separate theories of the strong and the electroweak interactions nor the conjectured unification of the two takes any account of gravity. Whether gravity can be described in a quantum theory and unified with the other fundamental forces remains an open question. Another basic deficiency of the standard model concerns the Higgs boson. The electroweak theory requires that the Higgs boson exist but does not specify precisely how the particle must interact with other particles or even with what its mass must be, except in the broadest terms.

THE SUPERCONDUCTING SUPER COLLIDER

What energy must we reach, and what new instruments do we need, to shed light on such fundamental problems? The questions surrounding the Higgs boson, although they are by no means the only challenges we face, are particularly well defined, and their answers will bear on the entire strategy of unification. They set a useful target for the next generation of machines.

It has been proposed that the Higgs boson is not an elementary particle at all but rather a composite object made up of elementary constituents analogous to quarks and leptons but subject to a new kind of strong interaction, often called technicolor, which would confine them within about 10^{-17} centimeter. The phenomena that would reveal such an interaction would become apparent at energies of about 1 TeV (trillion electron volts). A second approach to the question of the Higgs boson's mass and behavior employs a postulated principle known as supersymmetry, which relates particles that differ in spin. Supersymmetry entails the existence of an entirely new set of elusive, extremely massive particles. The new particles would correspond to known quarks, leptons and bosons but would differ in their spins. Because of their mass, such particles would reveal themselves fully only in interactions taking place at very high energy, probably about 1 TeV.

Our best hope for producing interactions of fundamental particles at energies of 1 TeV is an accelerator known as the Superconducting Super Collider (SSC), which is the subject of Chapter 5. Figure 1.9, prepared by James Freeman of the Collider Detector Group at the Fermi National Accelerator Laboratory (Fermilab) using the ISAJET model devised by Frank E. Paige, Jr., of the Brookhaven National Laboratory, illustrates how the SSC could shed light on the Higg's boson.

The SSC represents basic research at unprecedented cost on an unmatched scale. Yet the rewards will be proportionate. The advances of the past decade have brought us tantalizingly close to a profound new understanding of the fundamental constituents of nature and their interactions. Current theory suggests that the frontier of our ignorance falls at energies of about 1 TeV. Whatever clues about the unification of the forces of nature and the constituents of matter wait beyond that frontier, the SSC is likely to reveal them.

Quarks with Color and Flavor

The particles called quarks may be truly elementary. Their "colors" explain why they cannot be isolated; their "flavors" distinguish four basic kinds, including one that has the property called charm.

• • •

Sheldon Lee Glashow
October, 1975

tomos, the Greek root of "atom," means indivisible, and it was once thought that atoms were the ultimate, indivisible constituents of matter, that is, they were regarded as elementary particles. One of the principal achievements of physics in the 20th century has been the revelation that the atom is not indivisible or elementary at all but has a complex structure. In 1911 Ernest Rutherford showed that the atom consists of a small, dense nucleus surrounded by a cloud of electrons. It was subsequently revealed that the nucleus itself can be broken down into discrete particles, the protons and neutrons, and since then a great many related particles have been identified. During the past decade it has become apparent that those particles too are complex rather than elementary. They are now thought to be made up of the simpler things called quarks. A solitary quark has never been observed, in spite of many attempts to isolate one. Nonetheless, there are excellent grounds for believing they do exist. More important, quarks may be the last in the long series of progressively finer structures. They seem to be truly elementary.

When the quark hypothesis was first proposed more than 10 years ago, there were supposed to be three kinds of quark. The revised version of the theory I shall describe here requires 12 kinds. In the whimsical terminology that has evolved for the discussion of quarks they are said to come in four flavors, and each flavor is said to come in three colors. ("Flavor" and "color" are, of course, arbitrary labels; they have no relation to the usual meanings of those words.) One of the quark flavors is distinguished by the property called charm (another arbitrary term). The concept of charm was suggested in 1964, but until 1974 it had remained an untested conjecture. Several recent experimental findings, including the discovery in 1974 of the particles called *J* or psi, can be interpreted as supporting the charm hypothesis.

The basic notion that some subatomic particles are made of quarks has gained widespread acceptance, even in the absence of direct observational evidence. The more elaborate theory incorporating color and charm remains much more speculative. The views presented here are my own, and they are far from being accepted dogma. On the other hand, a growing body of evidence argues that these novel concepts must play some part in the description of nature. They help to bring together many seemingly unrelated theoretical developments of the past 15 years to form an elegant picture of the structure of matter. Indeed, quarks are at once the most rewarding and the most mystifying creation of modern particle physics. They are remarkably successful in explaining the structure of subatomic particles, but we cannot yet understand why they should be so successful.

The particles thought to be made up of quarks form the class called the hadrons. They are the only particles that interact through the "strong" force. Included are the protons and neutrons, and indeed it is the strong force that binds protons and neutrons together to form atomic nuclei. The strong force is also responsible for the rapid decay of many hadrons.

Another class of particles, defined in distinction to the hadrons, are the leptons. There are just four of them: the electron and the electron neutrino and muon and the muon neutrino (and their four antiparticles). The leptons are not subject to the strong force. Because the electron and the muon bear an electric charge, they "feel" the electromagnetic force, which is roughly 100 times weaker than the strong force. The two kinds of neutrino, which have no electric charge, feel neither the strong force nor the electromagnetic force, but interact solely through a third kind of force, weaker by several orders of magnitude and called the weak force. The strong force, the electromagnetic force and the weak force, together with gravitation, are believed to account for all interactions of matter. (Figure 1.1 contains an augmented table of quarks and leptons.)

The leptons give every indication of being elementary particles. The electron, for example, behaves as a point charge, and even when it is probed at the energies of the largest particle accelerators, no internal structure can be detected. The hadrons, on the other hand, seem complex. They have a measurable size: about 10^{-13} centimeter. Moreover, there are hundreds of them, all but a handful discovered in the past 25 years. Finally, all the hadrons, with the significant exception of the proton and the antiproton, are unstable in isolation. They decay into stable particles such as protons, electrons, neutrinos or photons. (The photon, which is the carrier of the electromagnetic force, is in a category apart; it is neither a lepton nor a hadron.)

The hadrons are subdivided into three families: baryons, antibaryons and mesons (see Figure 2.1).

The baryons include the proton and the neutron; the mesons include such particles as the pion. Baryons can be neither created nor destroyed except as pairs of baryons and antibaryons. This principle defines a conservation law, and it can be treated most conveniently in the system of bookkeeping that assigns simple numerical values, called quantum numbers, to conserved properties. In this case the quantum number is called baryon number. For baryons it is +1, for antibaryons −1 and for mesons 0. The conservation of baryon number then reduces to the rule that in any interaction the sum of the baryon numbers cannot change.

Baryon number provides a means of distinguishing baryons from mesons, but it is an artificial means, and it tells us nothing about the properties of the two kinds of particle. A more meaningful distinction can be established by examining another quantum number, spin angular momentum.

Under the rules of quantum mechanics a particle or a system of particles can assume only certain specified states of rotation, and hence can have only discrete values of angular momentum. The angular momentum is measured in units of $h/2\pi$, where h is Planck's constant, equal to about 6.6×10^{-27} ergsecond. Baryons are particles with a spin angular momentum measured in half-integral units, that is, with values of half an odd integer, such as 1/2 or 3/2. Mesons have integral values of spin angular momentum, such as 0 or 1.

The difference in spin angular momentum has important consequences for the behavior of the two kinds of hadron. Particles with integral spin are said to obey Bose-Einstein statistics (and are therefore called bosons). Those with half-integral spin obey Fermi-Dirac statistics (and are called fermions). In this context "statistics" refers to the behavior of a population of identical particles. Those that obey Bose-Einstein statistics can be brought together without restriction; an unlimited number of pions, for example, can occupy the same state. The Fermi-Dirac statistics, on the other hand, require that no two particles within a given system have the same energy and be identical in all their quantum numbers. This statement is equivalent to the exclusion principle formulated in 1925 by Wolfgang Pauli. He applied it in particular to electrons, which have a spin of 1/2 and are therefore fermions. It requires that each energy level in an atom contain only two electrons, with their spins aligned in opposite directions.

One of the clues to the complex nature of the

hadrons is that there are so many of them. Much of the endeavor to understand them has consisted of a search for some ordering principle that would make sense of the multitude.

The hadrons were first organized into small families of particles called charge multiplets or isotopic-spin multiplets; each multiplet consists of particles that have approximately the same mass and are identical in all their other properties except electric charge. The multiplets have one, two, three or four members. The proton and the neutron compose a multiplet of two (a doublet); both are considered to be manifestations of a single state of matter, the nucleon, with an average mass equivalent to an energy of .939 GeV (billion electron volts). The pion is a triplet with an average mass of .137 GeV and three charge states: +1,0 and −1. In the strong interactions the members of a multiplet are all equivalent, since electric charge plays no role in the strong interactions.

In 1962 a grander order was revealed when the charge multiplets were organized into "supermultiplets" that revealed relations between particles that differ in other properties in addition to charge. The creation of the supermultiplets was proposed independently by Murray Gell-Mann of the California Institute of Technology and by Yuval Ne'eman of Tel-Aviv University [see "Strongly Interacting Particles," by Geoffrey F. Chew, Murray Gell-Mann and Arthur H. Rosenfeld; SCIENTIFIC AMERICAN, February, 1964]. The introduction of the new system led directly to the quark hypothesis.

The grouping of the hadrons into supermultiplets involves eight quantum numbers and has been referred to as the "eightfold way." Its mathematical basis is a branch of group theory invented in the 19th century by the Norwegian mathematician Sophus Lie. The Lie group that generates the eightfold way is called $SU(3)$, which stands for the special unitary group of matrices of size 3×3. The theory requires that all hadrons belong to families corresponding to representations of the group $SU(3)$. The families can have one, three, six, eight, 10 or more members. If the eightfold way were an exact theory, all the members of a given family would have the same mass. The eightfold way is only an approximation, however, and within the families there are significant differences in mass.

The construction of the eightfold way begins with the classification of the hadrons into broad families sharing a common value of spin angular momentum. Each family of particles with identical spin is then depicted by plotting the distribution of two more quantum numbers: isotopic spin and strangeness.

Isotopic spin has nothing to do with the spin of a particle; it was given its name because it shares certain algebraic properties with the spin quantum number. It is a measure of the number of particles in a multiplet, and it is calculated according to the formula that the number of particles in the multiplet is one more than twice the isotopic spin. Thus the nucleon (a doublet) has an isotopic spin of 1/2; for the pion triplet the isotopic spin is 1.

Strangeness is a quantum number introduced to

	NAME	SYMBOL	MASS (GeV)	CHARGE STATES	BARYON NUMBER	SPIN	ISOTOPIC SPIN	STRANGENESS
BARYONS	NUCLEON	N	.939	0, +1	+1	$1/2$	$1/2$	0
	LAMBDA	Λ	1.115	0	+1	$1/2$	0	−1
	OMEGA	Ω	1.672	−1	+1	$3/2$	0	−2
MESONS	PION	π	.139	−1, 0, +1	0	0	1	0
	K	K	.496	0, +1	0	0	$1/2$	+1
	PHI	ϕ	1.019	0	0	1	0	0
	J	J	3.095	0	0	1	0	0

Figure 2.1 HADRONS are divided into baryons, made of three quarks, and mesons, made of a quark and an antiquark. (Antibaryons consist of three antiquarks.) The groups are distinguished by baryon number and by spin angular momentum, which has half-integral values for baryons and integral values for mesons. Each line in the table represents a multiplet of particles identical in all properties except electric charge, provided that small differences in mass are ignored. Isotopic spin is a function of the number of particles in a multiplet, and strangeness measures distribution of electric charge among them. Only a few representative hadrons are shown.

describe certain hadrons first observed in the 1950's and called strange particles because of their anomalously long lifetimes. They generally decay in from 10^{-10} to 10^{-7} second. Although that is a brief interval by everyday standards, it is much longer than the lifetime of 10^{-23} second characteristic of many other hadrons.

Like isotopic spin, strangeness depends on the properties of the multiplet, but it measures the distribution of charge among the particles rather than their number. The strangeness quantum number is equal to twice the average charge (the sum of the charges divided by the number of particles in a multiplet) minus the baryon number. By this contrivance it is made to vanish for all hadrons except the strange ones. The triplet of pions, for example, has an average charge of 0 and a baryon number of 0; its strangeness is therefore also 0. The nucleon doublet has an average charge of $+1/2$ and a baryon number of $+1$, so that those particles too have a strangeness of 0. On the other hand, the lambda particle is a neutral baryon that forms a family of one (a singlet). Its average charge of 0 and its baryon number of $+1$ give it a strangeness of -1.

On a graph that plots electric charge against strangeness the hadrons form orderly arrays. The mesons with a spin angular momentum of 0 compose an octet and a singlet; the octet is represented graphically as a hexagon with a particle at each vertex and two particles in the center, and the singlet is represented as a point at the origin. The mesons with a spin of 1 form an identical representation, and so do the baryons with a spin of 1/2. Finally, the baryons with a spin of 3/2 form a decimet (a group of 10) that can be graphed as a large triangle made up of a singlet, a doublet, a triplet and a quartet. The eightfold way was initially greeted with some skepticism, but the discovery in 1964 of the negatively charged omega particle, the predicted singlet in the baryon decimet, made converts of us all.

The regularity and economy of the supermultiplets are aesthetically satisfying, but they are also somewhat mystifying. The known hadrons do fit into such families, without exception. Mesons come only in families of one and eight, and baryons come only in families of one, eight and 10. The singlet, octet and decimet, however, are only a few of many possible representations of $SU(3)$. Families of three particles or six particles are entirely plausible, but they are not observed. Indeed, the variety of possible families is in principle infinite. Why, then, do only three representations appear in nature? It early became apparent that the eightfold way is in some approximate sense true, but it was also plain from the start that there is more to the story.

In 1963 an explanation was proposed independently by Gell-Mann and by George Zweig, also of Cal Tech. They perceived that the unexpected regularities could be understood if all hadrons were constructed from more fundamental constituents, which Gell-Mann named quarks. The quarks were to belong to the simplest nontrivial family of the eightfold way: a family of three. (There is also, of course, another family of three antiquarks.)

The quarks are required to have rather peculiar properties. Principal among these is their electric charge. All observed particles, without exception, bear integer multiples of the electron's charge; quarks, however, must have charges that are fractions of the electron's charge. Gell-Mann designated the three quarks u, d and s, for the arbitrary labels "up," "down" and "sideways."

The mechanics of the original quark model are completely specified by three simple rules. Mesons are invariably made of one quark and one antiquark. Baryons are invariably made of three quarks and antibaryons of three antiquarks. No other assemblage of quarks can exist as a hadron. The combinations of the three quarks under these rules are sufficient to account for all the hadrons that had been observed or predicted at the time. Furthermore, every allowed combination of quarks yields a known particle.

Many of the necessary properties of the quarks can be deduced from these rules. It is mandatory, for example, that each of the quarks be assigned a baryon number of $+1/3$ and each of the antiquarks a baryon number of $-1/3$. In that way any aggregate of three quarks has a total baryon number of $+1$ and hence defines a baryon; three antiquarks yield a particle with a baryon number of -1, an antibaryon. For mesons the baryon numbers of the quarks ($+1/3$ and $-1/3$) cancel, so that the meson, as required, has a baryon number of 0.

In a similar way the angular momentum of the hadrons is described by giving the quarks half-integral units of spin. A particle made of an odd number of quarks, such as a baryon, must therefore also have half-integral spin, conforming to the known characteristics of baryons. A particle made of an even number of quarks, such as a meson, must have integral spin.

The u quark and the s quark compose an isotopic-spin doublet: they have nearly the same mass and they are identical in all other properties except electric charge. The u quark is assigned a charge of $+2/3$ and the d quark is assigned a charge of $-1/3$. The average charge of the doublet is therefore $+1/6$ and twice the average charge is $+1/3$; since the baryon number of all quarks is $+1/3$, the definition of strangeness gives both the u and the d quarks a strangeness of 0. The s quark has a larger mass than either the u or the d and makes up an isotopic-spin singlet. It is given an electric charge of $-1/3$ and consequently has a strangeness of -1. The anti-quarks, denoted by writing the quark symbol with a bar over it, have opposite properties. The $\bar{u}$ has a charge of $-2/3$ and the $\bar{d}$ $+1/3$; both have zero strangeness. The $\bar{s}$ antiquark has a charge of $+1/3$ and a strangeness of $+1$.

Just two of the quarks, the u and the d, suffice to explain the structure of all the hadrons encountered in ordinary matter. The proton, for example, can be described by assembling two u quarks and a d quark; its composition is written uud. A quick accounting will show that all the properties of the proton determined by its quark constitution are in accord with the measured values. Its charge is equal to $2/3 + 2/3 - 1/3$, or $+1$. Similarly, its baryon number can be shown to be $+1$ and its spin $1/2$. A positive pion is composed of a u quark and a $\bar{d}$ antiquark (written $u\bar{d}$). Its charge is $2/3 + 1/3$, or $+1$; its spin and baryon number are both 0.

The third quark, s, is needed only to construct strange particles, and indeed it provides an explicit definition of strangeness: A strange particle is one that contains at least one s quark or $\bar{s}$ anti-quark. The lambda baryon, for example, can be shown from the charge distribution of its multiplet to have a strangeness of -1; that result is confirmed by its quark constitution of uds. Similarly, the neutral K meson, a strange particle, has a strangeness of $+1$, as confirmed by its composition of $d\bar{s}$.

Until quite recently these three kinds of quark were sufficient to describe all the known hadrons. As we shall see, experiments conducted during the past year seem to have created hadrons whose properties cannot be explained in terms of the original three quarks. The experiments can be interpreted as implying the existence of a fourth kind of quark, called the charmed quark and designated c.

The statement that the u, d and s quarks are sufficient to construct all the observed hadrons can be made more precisely in the mathematical formalism of the eightfold way. Since a meson is made up of one quark and one antiquark, and since there are three kinds, or flavors, of quark, there are nine possible combinations of quarks and antiquarks that can form a meson. It can be shown that one of these combinations represents a singlet and the remaining eight form an octet. Similarly, since a baryon is made up of three quarks, there are 27 possible combinations of quarks that can make up a baryon. They can be broken up into a singlet, two octets and a decimet. Those groupings correspond exactly to the observed families of hadrons. The quark theory thus explains why only a few of the possible representations of $SU(3)$ are realized in nature as hadron supermultiplets.

The quark rules provide a remarkably economical explanation for the formation of the observed hadron families. What principles, however, can explain the quark rules, which seem quite arbitrary? Why is it possible to bind together three quarks but not two or four? Why can we not create a single quark in isolation? A line of thought that leads to possible answers to these questions appeared at first as a defect in the quark theory.

As we have seen, it is necessary that the quarks have half-integral values of spin angular momentum; otherwise the known spins of the baryons and mesons would be predicted wrongly. Particles with half-integral spin are expected to obey Fermi-Dirac statistics and are therefore subject to the Pauli exclusion principle: No two particles within a particular system can have exactly the same quantum numbers. Quarks, however, seem to violate the principle. In making up a baryon it is often necessary that two identical quarks occupy the same state. The omega particle, for example, is made up of three s quarks, and all three must be in precisely the same state. That is possible only for particles that obey Bose-Einstein statistics. We are at an impasse: quarks must have half-integral spin but they must satisfy the statistics appropriate to particles having integral spin.

The connection between spin and statistics is an unshakable tenet of relativistic quantum mechanics. It can be deduced directly from the theory, and a violation has never been discovered. Since it holds for all other known particles, quarks could not reasonably be excluded from its dominion.

The concept that has proved essential to the solution of the quark statistics problem was proposed in 1964 by Oscar W. Greenberg of the University of

Maryland. He suggested that each flavor of quark comes in three varieties, identical in mass, spin, electric charge and all other measurable quantities but different in an additional property, which has come to be known as color. The exclusion principle could then be satisfied, and quarks could remain fermions, because the quarks in a baryon would not all occupy the same state. The quarks could differ in color even if they were the same in all other respects.

The color hypothesis requires two additional quark rules. The first simply restates the condition that color was introduced to satisfy: Baryons must be made up of three quarks, all of which have different colors. The second describes the application of color to mesons: Mesons are made of a quark and an antiquark of the same color, but with equal representation of each of the three colors. The effect of these rules is that no hadron can exhibit net color. A baryon invariably contains quarks of each of the three colors, say red, yellow and blue. In the meson one can imagine the quark and antiquark as being a single color at any given moment, but continually and simultaneously changing color, so that over any measurable interval they will both spend equal amounts of time as red, blue and yellow quarks.

The price of the color hypothesis is a tripling of the number of quarks; there must be nine instead of three (with charm yet to be considered). At first it may also appear that we have greatly increased the number of hadrons, but that is an illusion. With color there seem to be nine times as many mesons and 27 times as many baryons, but the rules for assembling hadrons from colored quarks ensure that none of the additional particles are observable.

Although the quark rules imply that we will never see a colored particle, the color hypothesis is not merely a formal construct without predictive value. The increase it requires in the number of quarks can be detected in at least two ways. One is through the effect of color on the lifetime of the neutral pion, which almost always decays into two photons. Stephen L. Adler of the Institute for Advanced Study has shown that its rate of decay depends on the square of the number of quark colors. Just the observed lifetime is obtained by assuming that there are three colors.

Another effect of color can be detected in experiments in which electrons and their antiparticles, the positrons, annihilate each other at high energy. The outcome of such an event is sometimes a group of hadrons and sometimes a muon and an antimuon.

At sufficiently high energy the ratio of the number of hadrons to the number of muon-antimuon pairs is expected to approach a constant value, equal to the sum of the squares of the charges of the quarks. Tripling the number of quarks also triples the expected value of the ratio. The experimental result at energies of from 2 GeV to 3 GeV is in reasonable agreement with the color hypothesis (which predicts a value of 2) and is quite incompatible with the original theory of quarks without color.

The introduction of the color quantum number solves the problem of quark statistics, but it once again requires a set of rules that seem arbitrary. The rules can be accounted for, however, by establishing another hypothetical symmetry group analogous to the $SU(3)$ symmetry proposed by Gell-Mann and by Ne'eman. The earlier $SU(3)$ is concerned entirely with combinations of the three quark flavors; the new one deals exclusively with the three quark colors. Moreover, unlike the earlier theory, which is only approximate, color $SU(3)$ is supposed to be an exact symmetry, so that quarks of the same flavor but different color will have identical masses.

In the color $SU(3)$ theory all the quark rules can be explained if we accept one postulate: All hadrons must be represented by color singlets; no larger multiplets can be allowed. A color singlet can be constructed in two ways: by combining an identically colored quark and antiquark with all three colors equally represented, or by combining three quarks or three antiquarks in such a way that the three colors are all included. These conditions, of course, are equivalent to the rules for building mesons, baryons and antibaryons, and they ensure that all hadrons will be colorless. There are no other ways to make a singlet in color $SU(3)$; a particle made any other way would be a member of a larger multiplet, and it would display a particular color.

Although the color $SU(3)$ theory of the hadrons can explain the quark rules, it cannot entirely eliminate the arbitrary element in their nature. We can ask a still more fundamental question: What explains the postulate that all hadrons must be color singlets? One approach to an answer, admittedly a speculative one, has been suggested recently by many investigators; it incorporates the color $SU(3)$ model of the hadrons into one of the class of theories called gauge theories.

The color gauge theory postulates the existence of eight massless particles, sometimes called gluons, that are the carriers of the strong force, just as the

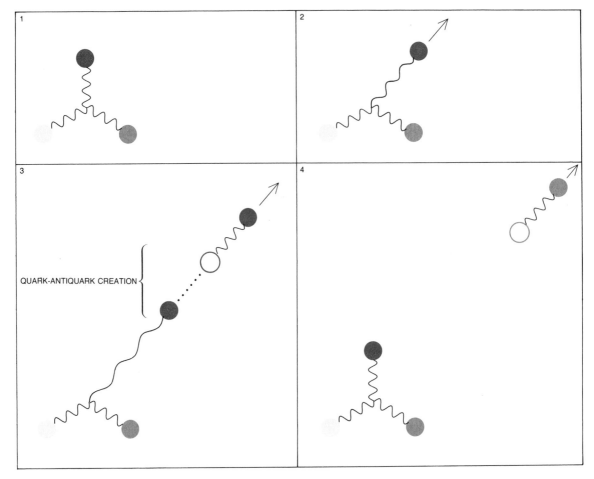

Figure 2.2 CONFINEMENT OF QUARKS can be explained if the quarks are assumed to come in three colors. A baryon is a bound system of three quarks, including one of each of the three colors (1), so that the baryon as a whole is colorless. The strong force between colored quarks differs in character from that between colorless composite particles: it does not diminish with distance but remains constant. As a result, when a quark is separated from a baryon (2), the potential energy of the system rapidly increases and a quark and an antiquark are created (3). The new quark restores the baryon to its original form, and the antiquark (*open circle*) adheres to the dislodged quark, forming another kind of particle, a meson (4). At any one moment the quark and the antiquark in a meson are the same color, but the three colors are equally represented.

photon is the carrier of the electromagnetic force. Like the photon, they are electrically neutral, and they have a spin of 1; they are therefore called vector bosons (bosons because they have integer spin and obey Bose-Einstein statistics, vector because a particle with a spin of 1 is described by a wave function that takes the form of a four-dimensional vector). Gluons, like quarks, have not been detected.

When a quark emits or absorbs a gluon, the quark changes its color but not its flavor. For example, the emission of a gluon might transform a red u quark into a blue or a yellow u quark, but it could not change it into a d or an s quark of any color. Since the color gluons are the quanta of the strong force, it follows that color is the aspect of quarks that is most important in the strong interactions. In fact, when describing interactions that involve only the strong force, one can virtually ignore the flavors of quarks.

The color gauge theory proposes that the force

that binds together colored quarks represents the true character of the strong interaction. The more familiar strong interactions of hadrons (such as the binding of protons and neutrons in a nucleus) are manifestations of the same fundamental force, but the interactions of colorless hadrons are no more than a pale remnant of the underlying interaction between colored quarks. Just as the van der Waals force between molecules is only a feeble vestige of the electromagnetic force that binds electrons to nuclei, the strong force observed between hadrons is only a vestige of that operating within the individual hadron.

From these theoretical arguments one can derive an intriguing, if speculative, explanation of the confinement of quarks. It has been formulated by John Kogut and Kenneth Wilson of Cornell University and by Leonard Susskind of Yeshiva University. If it should be proved correct, it would show that the failure to observe colored particles (such as isolated quarks and gluons) is not the result of any experimental deficiency but is a direct consequence of the nature of the strong force.

The electromagnetic force between two charged particles is described by Coulomb's law: The force decreases as the square of the distance between the charges. Gravitation obeys a fundamentally similar law. At large distances both forces dwindle to insignificance. Kogut, Wilson and Susskind argue that the strong force between two colored quarks behaves quite differently: it does not diminish with distance but remains constant, independent of the separation of the quarks. If their argument is sound, an enormous amount of energy would be required to isolate a quark.

Separating an electron from the valence shell of an atom requires a few electron volts. Splitting an atomic nucleus requires a few million electron volts. In contrast to these values, the separation of a single quark by just an inch from the proton of which it is a constituent would require the investment of 10^{13} GeV, enough energy to separate the author from the earth by some 30 feet. Long before such an energy level could be attained another process would intervene. From the energy supplied in the effort to extract a single quark, a new quark and antiquark would materialize. The new quark would replace the one removed from the proton, and would reconstitute that particle. The new antiquark would adhere to the dislodged quark, making a meson. Instead of isolating a colored quark, all that is accomplished is the creation of a colorless meson as

seen in Figure 2.2. By this mechanism we are prohibited from ever seeing a solitary quark or a gluon or any combination of quarks or gluons that exhibits color.

If this interpretation of quark confinement is correct, it suggests an ingenious way to terminate the apparently infinite regression of finer structures in matter. Atoms can be analyzed into electrons and nuclei, nuclei into protons and neutrons, and protons and neutrons into quarks, but the theory of quark confinement suggests that the series stops there. It is difficult to imagine how a particle could have an internal structure if the particle cannot even be created.

Quarks of the same flavor but different color are expected to be identical in all properties except color; indeed, that is why the concept of color was introduced. Quarks that differ in flavor, however, have quite different properties. It is because the u quark and the d quark differ in electric charge that the proton is charged and the neutron is not. Similarly, it is because the s quark is considerably more massive than either the u or the d quark that strange particles are generally the heaviest members of their families. The charmed quark, c, must be heavier still, and charmed particles as a rule should therefore be heavier than all others. It is the flavor of quarks that brings variety to the world of hadrons, not their color.

As we have seen, the flavors of quarks are unaffected by the strong interactions. In a weak interaction, on the other hand, a quark can change its flavor (but not its color). The weak interactions also couple quarks to the leptons. The classical example of this coupling is nuclear beta decay, in which a neutron is converted into a proton with the emission of an electron and an antineutrino. In terms of quarks the reaction represents the conversion of a d quark to a u quark, accompanied by the emission of the two leptons.

The weak interactions are thought to be mediated by vector bosons, just as the strong and the electromagnetic interactions are. The principal one, labeled W and long called the intermediate vector boson, was predicted in 1938 by Hideki Yukawa. It has an electric charge of -1, and it differs from the photon and the color gluons in that it has mass, indeed a quite large mass. Quarks can change their flavor by emitting or absorbing a W particle. Beta decay, for example, is interpreted as the emission of a W by a d quark, which converts the quark into a u; the W then decays to yield the electron and antineutrino. From this process it follows that the W can also interact with leptons, and it thus provides a link between the two groups of apparently elementary particles.

The realization that the strong, weak and electromagnetic forces are all carried by the same kind of particle — bosons with a spin of 1 – invites speculation that all three might have a common basis in some simple unified theory. A step toward such a unification would be the reconciliation of the weak interactions and electromagnetism. Julian Schwinger of Harvard University attempted such a unification in the mid-1950's (when I was one of his doctoral students, working on these very questions). His theory had serious flaws. One was eliminated in 1961, when I introduced a second, neutral vector boson, now called Z, to complement the electrically charged W. Other difficulties persisted for 10 years, until in 1967 Steven Weinberg of Harvard and Abdus Salam of the International Center for Theoretical Physics in Trieste independently suggested a resolution. By 1971 it was generally agreed, largely because of the work of Gerard 't Hooft of the University of Utrecht, that the Weinberg-Salam conjecture is successful [see "Unified Theories of Elementary-Particle Interaction," by Steven Weinberg; SCIENTIFIC AMERICAN, July, 1974].

Through the unified weak and electromagnetic interactions, quarks and leptons are intimately related. These interactions "see" the four leptons and distinguish between the three quark flavors. The W particle can induce one kind of neutrino to become an electron and the other kind of neutrino to become a muon. Similarly, the W can convert a u quark into a d quark; it can also influence the u quark to become an s quark, although much less readily.

There is an obvious lack of symmetry in these relations. The leptons consist of two couples, married to each other by the weak interaction: the electron with the electron neutrino and the muon with the muon neutrino. The quarks, on the other hand, come in only three flavors, and so one must remain unwed. The scheme could be made much tidier if there were a fourth quark flavor, in order to provide a partner for the unwed quark. Both the quarks and the leptons would then consist of two pairs of particles, and each member of a pair could change into the other member of the same pair simply by emitting a W. The desirability of such lepton-quark symmetry led James Bjorken and me, among others, to

postulate the existence of a fourth quark in 1964. Bjorken and I called it the charmed quark. When provisions are made for quark colors, charm becomes a fourth quark flavor, and a new triplet of colored quarks is required. There are thus a total of 12 quarks.

Since 1964 several additional arguments for charm have developed. To me the most compelling of them is the need to explain the suppression of certain interactions called strangeness-changing neutral currents. An explanation that relies on the properties of the charmed quark was presented in 1967 by John Iliopoulos, Luciano Maiani and me.

Strangeness-changing neutral currents are weak interactions in which the net electric charge of the hadrons does not change but the strangeness does; typically an s quark is transformed into a d quark, and two leptons are emitted. An example is the decay of the neutral K meson (a strange particle) into two oppositely charged muons. Such processes are found by experiment to be extremely rare. The three-quark theory cannot account for their suppression, and in fact the unified theory of weak and electromagnetic interactions predicts rates more than a million times greater than those observed.

The addition of a fourth quark flavor with the same electric charge as the u quark neatly accounts for the suppression, although the mechanism by which it does so may seem bizarre. With two pairs of quarks there are two possible paths for the strangeness-changing interactions, instead of just one when there are only three quarks. In the macroscopic world the addition of a second path, or channel, would be expected always to bring an increase in the reaction rate. In a world governed by quantum mechanics, however, it is possible to subtract as well as to add. As it happens, a sign in the equation that defines one of the reactions is negative, and the two interactions cancel each other.

The addition of a fourth quark flavor must obviously increase the number of hadrons. In order to accommodate the newly predicted particles in supermultiplets the eightfold way must be expanded. In particular another dimension must be added to the graphs employed to represent the families, so that the plane figures of the earlier symmetry become Platonic and Archimedean solids.

To the meson octet are added six charmed particles and one uncharmed particle to make up a new family of 15. It is represented as a cuboctahedron, in which one plane contains the hexagon of the original uncharmed meson octet. The baryon octets and decimet are expected to form two families having 20 members each. They are represented as a tetrahedron truncated at each vertex and as a regular tetrahedron. In addition there is a smaller regular tetrahedron consisting of just four baryons. Again, each figure contains one plane of uncharmed particles seen in Figure 2.3.

It now appears that the first of the new particles to be discovered is a meson that is not itself charmed. That conclusion is based on the assumption that the predicted meson is the same particle as the *J* or psi particle discovered in November 1974. The announcement of the discovery was made simultaneously by Samuel C. C. Ting and his colleagues at the Brookhaven National Laboratory and by Burton Richter, Jr., and a group of other physicists at the Stanford Linear Accelerator Center (SLAC). At Brookhaven it was named *J*, at Stanford psi. Here I shall adopt the name *J*. For two excited states of the same particle, however, the names psi' and psi" will be employed, since they were seen only in the SLAC experiments.

The *J* particle was found as a resonance, an enhancement at a particular energy in the probability of an interaction between other particles. At Brookhaven the resonance was detected in the number of electron-positron pairs produced in collisions between protons and atomic nuclei. At SLAC it was observed in the products of annihilations of electrons and positrons (see Figure 2.4). The energy at which the resonances were observed — and thus the energy or mass of the *J* particle — is about 3.1 GeV [see "Electron-Positron Annihilation and the New Particles," by Sidney D. Drell; SCIENTIFIC AMERICAN, June, 1975].

The *J* particle decays in about 10^{-20} second, certainly a brief interval, but nevertheless 1,000 times longer than the expected lifetime of a particle having the *J*'s mass. The considerable excitement generated by the discovery of the *J* was largely a result of its long lifetime.

A great many explanations of the particle were proposed; for example, it was suggested that it might be the Z. I believe there is good reason to interpret the *J* as being a meson made up of a charmed quark and a charmed antiquark, that is, a meson with the quark constitution $c\bar{c}$ (see Figure 2.3). Thomas Applequist and H. David Politzer of Harvard have named such a meson "charmonium," by analogy to positronium, a bound state of an electron and a positron. Charmonium is without

charm because the charm quantum numbers of its quarks (+1 and −1) add up to zero.

The charmonium hypothesis can account for the anomalous lifetime of the *J* if one considers the ultimate fate of the decaying particle's quarks. There are three possibilities: they can be split up to become constituents of two daughter hadrons, they can both become part of a single daughter particle or they can be annihilated. An empirical rule, first noted by Zweig, states that decays of the first kind are allowed but the other two are suppressed. For the *J* particle to decay in the allowed manner it must create two charmed particles, that is, two hadrons, one containing a charmed quark and the other a charmed antiquark. That decay is possible only if the mass of the *J* is greater than the combined masses of the charmed daughter particles. There is reason to believe the lightest charmed particle has a mass greater than half of the mass of the *J*, and it therefore appears that the *J* cannot decay in the allowed mode. The *J* cannot decay in the second way, either, keeping both its quarks in a single particle, because the *J* is the least massive state containing a charmed quark and a charmed antiquark. It must therefore decay by the annihilation of its quarks, a decay suppressed by Zweig's rule. The suppression offers a partial explanation for the particle's extended lifetime.

Zweig's rule was formulated to explain the decay of the phi meson, which is made up of a strange quark and a strange antiquark and has a mass of about 1 GeV. The two particles are closely analogous, but the decay of the *J* is appreciably slower than that of the phi. Why should Zweig's rule be more effective for *J* than it is for phi? Furthermore, what explains Zweig's arbitrary rule?

A possible answer is provided by the theoretical concept called asymptotic freedom, which holds that the strong interactions become less strong at high energy. At sufficiently high energy the proton behaves as if it were made up of three freely moving quarks instead of three tightly bound ones. The concept takes its name from the fact that the quarks approach the state of free motion asymptotically as the energy is increased. Asymptotic freedom offers an explanation for the discrepancy between the phi and the *J* particles in the application of Zweig's rule. Because the *J* is so massive, or alternatively so energetic, the strong interaction is of diminished strength, and it is particularly difficult for the quark and the antiquark to annihilate each other.

Like positronium, charmonium should appear in

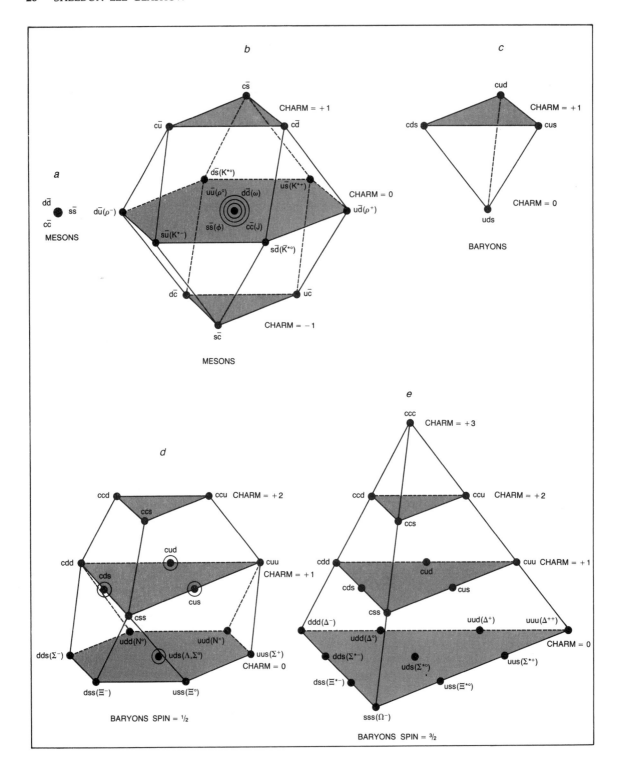

Figure 2.3 SUPERMULTIPLETS OF HADRONS arranged as polyhedrons. Each supermultiplet consists of particles with the same spin angular momentum. Particles are assigned positions according to three quantum numbers: positions on shaded planes are determined by isotopic spin and strangeness; the planes themselves indicate charm. Mesons (here spin 1) are represented by a point (*a*) and by a cuboctahedron (*b*), which comprises 15 particles, including six charmed ones. Baryons form a small regular tetrahedron (*c*) of four particles, a truncated tetrahedron (*d*) of 20 particles and a large regular tetrahedron (*e*) of 20 particles. For those particles that have been observed the established symbol is also given. Each figure contains one plane (*color*) of uncharmed particles.

many energy states. Two were discovered at SLAC soon after the first state was found; they are psi', with a mass of about 3.7 GeV, and psi'', with a mass of about 4.1 GeV. They appear to be simple excited states of the lowest-lying state of charmonium, the J particle. Psi' decays only a little more quickly than J, and half the time its decay products are the J particle itself and two pions. Thus it sometimes decays by the second suppressed process described by Zweig's rule, that is, by contributing both of its quarks to a single daughter particle. The extended

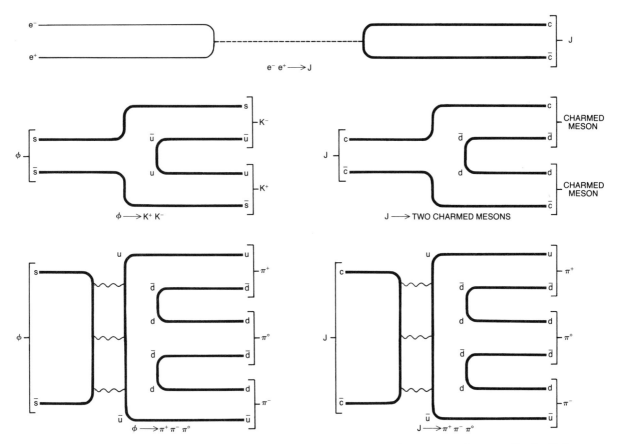

Figure 2.4 J PARTICLE is interpreted as a bound state called charmonium. It is made when electrons and positrons annihilate each other (*top*) to create a virtual photon (*broken line*), which yields the new meson. The preferred decay mode for both the J and phi meson consists in contributing the quark and the antiquark to two daughter mesons. For the phi meson that is just possible (*middle left*) because the mass of the phi is slightly greater than the combined mass of two strange K mesons. For the J it is not possible (*middle right*) because the lightest charmed particle is more than half as massive as the J. The J must decay by the annihilation of its quarks (*bottom right*), which yields three gluons (*wavy lines*) that are transformed into three pions. This mode of decay is suppressed, and the equivalent process for the phi meson (*bottom left*) is rarely observed.

lifetime implies that psi' also lies below the energy threshold for the creation of a pair of charmed particles.

Psi" decays much more quickly and therefore must be decaying in some mode permitted by Zweig's rule. Its decay products have not yet been determined, but it is possible they include charmed hadrons.

Numerous other excited states of charmonium follow inevitably from the theory of quark interactions (see Figure 9.7, right). One, called *p*-wave charmonium, is formed when the particle takes on an additional unit of angular momentum. Some fraction of the time psi' should decay into *p*-wave charmonium, which should subsequently decay

predominantly to the ground state, *J*. At each transition a photon of characteristic energy must be emitted. Recent experiments at the DORIS particle-storage rings of the German Electron Synchrotron in Hamburg have apparently detected the decays associated with the *p*-wave particle. In a few percent of its decays psi' yields the *J* particle and two photons, with energies of .2 GeV and .4 GeV. At SLAC psi' has been found to decay into an intermediate state and a single photon with an energy of .2 GeV. The intermediate state, which is presumably the same particle as the one observed at DORIS, then decays directly into hadrons. (This is discussed further in Chapter 9.)

The correspondence of theory and experiment re-

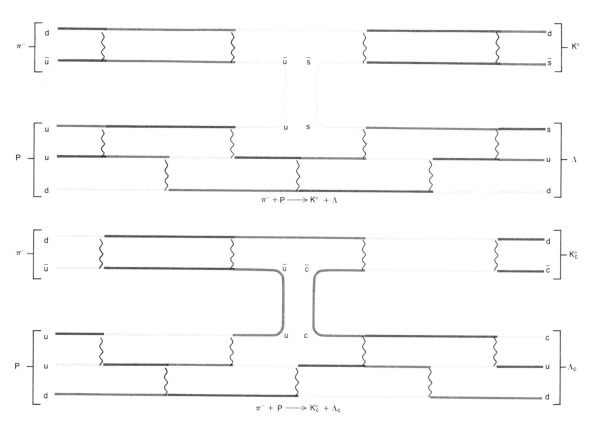

Figure 2.5 STRANGE AND CHARMED PARTICLES should be created in interactions of ordinary matter. The interactions are displayed here as intersections of lines representing quarks and other particles. Within a hadron the quarks continually exchange massless particles called gluons (*wavy lines*), the carriers of the strong force. By emitting a gluon a quark changes its color but not its flavor. Strange particles can be created (*top*) when a *u*

quark in a proton and a $\bar{u}$ antiquark in a pion annihilate each other and give rise to an *s* quark and an $\bar{s}$ antiquark. Products are a *K* meson and a lambda baryon. At higher energy the same annihilation could yield a *c* quark and a $\bar{c}$ antiquark (*bottom*). This process, which has not been observed, would yield a charmed meson and a charmed baryon.

vealed by the discovery of the *p*-wave transitions inspires considerable confidence that the charmonium interpretation of the *J* particle is correct. There is at least one more predicted state, called paracharmonium, that must be found if this explanation of the particle is to be confirmed. It differs from the observed states in the orientation of the quark spins: in *J*, psi' and psi" (collectively called orthocharmonium) they are parallel; in paracharmonium they are antiparallel. Paracharmonium has so far evaded de-

tection, but if the theoretical description is to make sense, paracharmonium must exist.

In addition to the various states of (uncharmed) charmonium, all the predicted charmed particles must also exist. If the *J* is in fact a state of charmonium, we can deduce from its mass the masses of all the hadrons containing charmed quarks (see Figure 2.6).

An important initial constraint on the range of

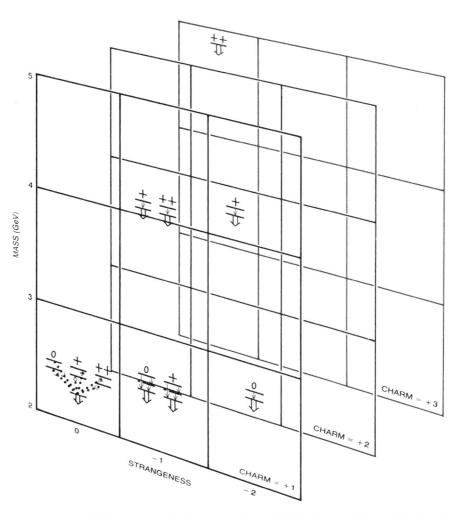

Figure 2.6 CHARMED BARYONS are expected to be considerably more massive than other hadrons. Some of the charmed particles must exist in more than one charge state (indicated by zeros and plus signs) and at several energy levels (indicated by their position with respect to the mass scale at left). Some of the particles can decay by the strong interaction (*dotted arrows*) or the electromagnetic interac-

tion (*solid black arrows*) into states that have the same quantum numbers but smaller mass; others can decay only by the weak interaction (*open arrows*) into uncharmed particles. The form of the table is determined largely by the requirement that a baryon be made up of exactly three quarks.

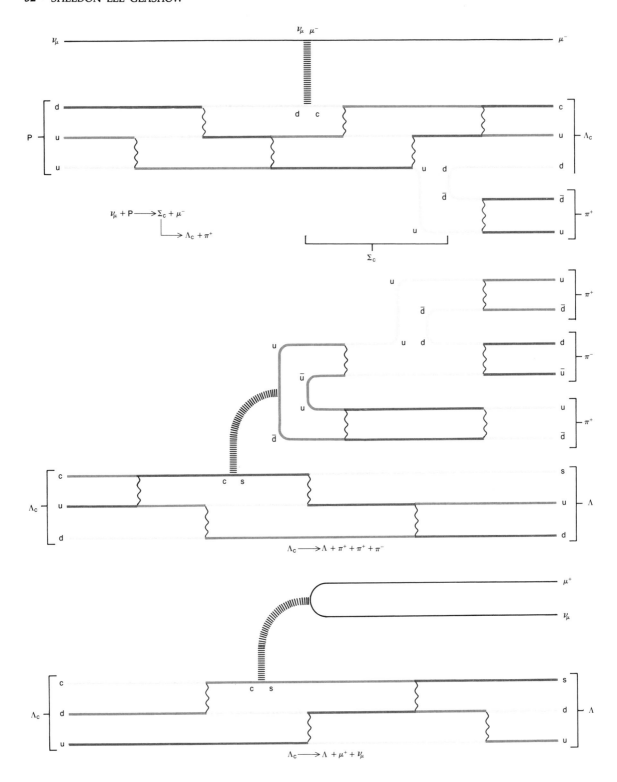

$\nu_\mu + P \longrightarrow \Sigma_c + \mu^-$
$ \Sigma_c \longrightarrow \Lambda_c + \pi^+$

$\Lambda_c \longrightarrow \Lambda + \pi^+ + \pi^+ + \pi^-$

$\Lambda_c \longrightarrow \Lambda + \mu^+ + \nu_\mu$

Figure 2.7 WEAK INTERACTIONS, mediated by the W particle (*hatched lines*), can change the flavor of a quark, but they have no effect on its color. Single charmed particles can therefore be created by the weak force. A neutrino interacts with a proton (*top*), converting a d quark into a c quark while the neutrino is transformed into a muon. The immediate product is a charmed sigma baryon, but that quickly decays by the emission of a pion to a charmed lambda baryon. The charmed lambda particle can itself decay by emitting a W particle, converting the c quark into an s quark and thereby forming a strange lambda baryon. The strange lambda baryon can be accompanied by three pions, produced by repeated quark-pair creation (*middle*), or by a muon and a neutrino (*bottom*).

possible masses was provided by the interpretation of the suppression of strangeness-changing neutral currents. If the suppression mechanism is to work, the charmed quark cannot be too much heavier than its siblings. On the other hand, it cannot be very light or charmed hadrons would already have been observed. An estimate from these conditions suggested that charmed particles would be found to have masses of about 2 or 3 GeV.

After the discovery of the J, I performed a more formal analysis with my colleagues at Harvard, Alvaro De Rújula and Howard Georgi. So did many others. Our estimates indicate that the least massive charmed states are mesons made up of a c quark and a $\bar{u}$ or $\bar{d}$ antiquark; their mass should fall between 1.8 GeV and 2.0 GeV. A value within that range could be in agreement with the supposition that psi′ lies below the threshold for the creation of a pair of charmed mesons, but psi″ lies above it.

The least massive charmed baryon has a quark composition of *udc*; we predict that its mass is near 2.2 GeV. As might be expected, since the c quark is the heaviest of the four, the most massive predicted charmed hadron is the *ccc* baryon. We estimate its mass at about 5 GeV.

An important principle guiding experimental searches for charmed hadrons is the requirement that in most kinds of interaction charmed particles can be created only in pairs. Two hadrons must be produced, one containing a charmed quark, the other a charmed antiquark; the obvious consequence is a doubling of the energy required to create a charmed particle. An important exception to this rule is the interaction of neutrinos with other kinds of particles, such as protons (see Figure 2.7). Neutrino events are exempt because neutrinos have only weak interactions and quark flavor can be changed in weak processes. Many experimental

techniques have been tried in the search for charm during the past 10 years, yet no charmed particle has been unambiguously identified. Nevertheless, two recent experiments, both involving neutrino interactions, are encouraging. In both charm may at last have appeared, but even if that proves to be an illusion, the experiments suggest promising lines of research.

One of the experiments was conducted at the Fermi National Accelerator Laboratory in Batavia, Ill., by a group of physicists headed by David B. Cline of the University of Wisconsin, Alfred K. Mann of the University of Pennsylvania and Carlo Rubbia of Harvard. In examining the interactions of high-energy neutrinos they found that in several percent of the events the products included two oppositely charged muons. One of the muons could be created directly from the incident neutrino, but the other is difficult to account for with only the ensemble of known, uncharmed particles. The most likely interpretation is that a heavy particle created in the reaction decays by the weak force to emit the muon. The particle would have a mass of between 2 and 4 GeV, and if it is a hadron, some explanation must be found for its weak decay. Most particles with masses that large decay by the strong force. The presence of a charmed quark in the particle might provide the required explanation.

The second experiment was performed at Brookhaven by a group of investigators under Nicholas P. Samios. They photographed the tracks resulting from the interaction of neutrinos with protons in a bubble chamber. In a sample of several hundred observed collisions one photograph seemed to have no conventional interpretation (see Figure 2.8). The final state can be construed as the decay products of a charmed baryon. The process would provide convincing evidence for the existence of charm if it were not attested to by only one event. A few more observations of the same reaction would settle the matter.

It would be misleading to give the impression that the description of hadrons in terms of quarks of three colors and four flavors has solved all the outstanding problems in the physics of elementary particles. For example, continuing measurements of the ratio of hadrons to muon pairs produced in electron-positron annihilations have confounded prediction. The ratio discriminates between various quark models, and an argument in support of the

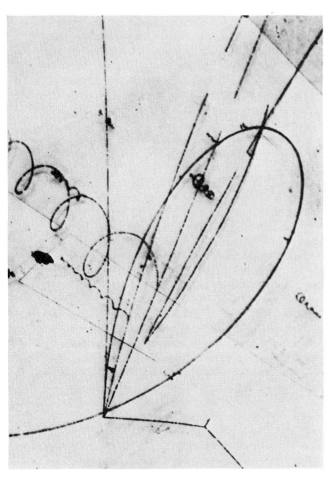

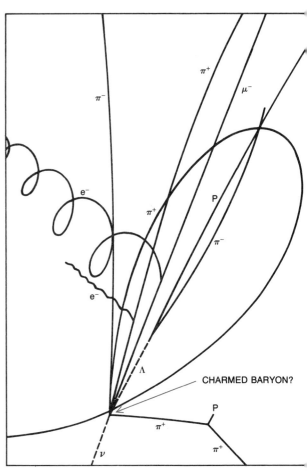

Figure 2.8 DETECTION OF A CHARMED BARYON from the collision of a neutrino and a proton? The photo at left was made in a liquid hydrogen bubble chamber; the particle tracks in the photo are identified in the diagram. The neutrino enters from bottom left, but its track is not visible. The possible charmed baryon does have an electric charge, but its track cannot be seen either because it is too short. The charmed particle decays into a neutral lambda particle, a strange baryon. The lambda particle leaves no track, but its decay products form a vertex that points toward the initial interaction. Four pions and a muon are also created, and two electrons struck by fast-moving particles spiral to the left in the bubble chamber's magnetic field. (Photo by N. P. Samios.)

color hypothesis was that at energies of from 2 to 3 GeV the ratio is about 2. At higher energy, high enough for charmed hadrons to be created in pairs, the ratio was expected to rise from 2 to about 3.3. The ratio does increase, but it overshoots the mark and appears to stabilize at a value of about 5. Perhaps charmed particles are being formed, but it seems that something else is happening as well: some particle is being made that does not appear in the theory I have described. One of my colleagues at Harvard, Michael Barnett, believes we have not

been ambitious enough. He invokes six quark flavors rather than four, so that there are three flavors of charmed quark. It is also possible there are heavier leptons we know nothing about.

Finally, even if a completely consistent and verifiable quark model could be devised, many fundamental questions would remain. One such perplexity is implicit in the quark-lepton symmetry that led to the charm hypothesis. Both the quarks and the leptons, all of them apparently elementary, can be divided into two subgroups. In one group are the u

and *d* quarks and the electron and electron neutrino. These four particles are the only ones needed to construct the world; they are sufficient to build all atoms and molecules, and even to keep the sun and other stars shining. The other subgroup consists of the strange and charmed quarks and the muon and muon neutrino. Some of them are seen occasionally in cosmic rays, but mainly they are made in high-energy particle accelerators. It would appear that nature could have made do with half as many fundamental things. Surely the second group was not created simply for the entertainment or edification of physicists, but what is the purpose of this grand doubling? At this point we have no answer.

The Lattice Theory of Quark Confinement

The force between quarks in a particle such as the proton has been simulated by imposing a discrete lattice on the structure of space and time. The results suggest why a free quark cannot be isolated.

. . .

Claudio Rebbi

February, 1983

The development of quantum mechanics put to rest the uncritical acceptance of the idea that elementary particles are the "building blocks" of matter. Often such particles do not act like hard, impenetrable blocks at all, and in many circumstances they must be described as waves. Until recently, however, it still seemed that elementary particles were like building blocks at least to the extent that each particle could in principle be isolated and observed as an individual entity. The electron, the proton and the neutron, for example, can be separated from one another and observed as individual packets of waves. Even this limited interpretation of the building-block metaphor fails in the case of the quark, the supposed constituent of the proton, the neutron and many related particles. Apparently a quark cannot be isolated; although there is abundant evidence for the existence of quarks and antiquarks bound together in pairs and triplets, an individual, or free, quark has never been observed.

As the experimental evidence has accumulated, it has begun to seem that if quarks are real particles at all, they must be permanently bound within nuclear particles. Any theory of quark interactions ought to account for this phenomenon, which is called quark confinement. It is easy to construct pictorial models of particles such as the proton in which the constituent quarks are confined. For example, the quarks can be thought of as being fastened to the ends of an unbreakable string; they are then free to move about within the volume defined by the length of the string but cannot wander away from one another. It is a formidable task, however, to formulate a theory that can account for the permanent binding of quarks and the structure of nuclear particles without violating the constraints imposed by the theory of relativity, quantum mechanics and the principles of ordinary causality.

After several years of both experimental and theoretical investigations most particle physicists are confident they at last have a theory capable of explaining the interactions of quarks. One reason for confidence is that the theory is a mathematical analogue of the most successful physical theory ever developed: the quantum theory of interactions in an electromagnetic field. The latter theory is called quantum electrodynamics, or QED, and the conceptual similarity of the theory of quark interactions to QED is reflected in the name of the new theory: quantum chromodynamics, or QCD.

The difficulty that has delayed full acceptance of QCD is that its mathematical complexity makes any rigorous, analytical prediction from it exceedingly difficult. Indeed, up to now the most eagerly sought prediction of QCD, namely the demonstration of quark confinement, has not been forthcoming. Recently, however, my colleagues and I at the Brookhaven National Laboratory have applied mathematical methods that rely heavily on the capabilities of the digital computer to the problem of confinement, and a numerical breakthrough has been achieved. Because the method explores the implications of QCD by making a series of increasingly accurate approximations, the results of the calculations do not carry the same force as a logical deduction from accepted first principles. Nevertheless, the numerical results have provided strong evidence for the confinement of quarks.

The framework for the calculations is a pioneering suggestion made in 1974 by Kenneth G. Wilson of Cornell University. Wilson proposed that QCD be formulated on a cubic lattice, an array that divides space and time into discrete points (see Figure 3.1). The lattice is only an approximation to real space-time, but it allows calculations to be made that would otherwise be impossible. As the mesh of the lattice is made progressively finer the values of physical quantities defined on the lattice converge to the values QCD would predict for them in ordinary, continuous space and time. Our numerical approximations show that for an extremely fine lattice, confinement is a consequence of QCD. For reasons that will become clear both QCD and QED are called gauge theories; the computational method I shall describe is therefore called a lattice gauge theory.

The original impetus for the quark model was the need to bring order to the large number of particles that exhibit strong interactions, or in other words those subject to the strong force. The proton and the neutron are members of this class, and indeed it is the strong force that binds them in an atomic nucleus. The existence of many other strongly interacting particles has been inferred from the decay products of collisions in accelerators. Most such particles live for an extremely short time,

as short as 10^{-24} second, before they decay into other particles. All particles that are subject to the strong force are called hadrons, from the Greek adjective hadros, meaning robust or heavily built.

In 1962 Murray Gell-Mann of the California Institute of Technology and Yuval Ne'eman of Tel-Aviv University proposed a scheme for classifying the hadrons in symmetrical patterns. The scheme was based on the mathematical theory of groups and was called the eightfold way. A short time afterward Gell-Mann and, independently, George Zweig, also of Cal Tech, proposed a physical interpretation of the eightfold way. The mathematical classification could be explained by assuming that all hadrons are built up of more fundamental constituents, which Gell-Mann called quarks.

At the time every known hadron could be understood as some combination of three basic quarks (and their corresponding antiquarks): the up or u quark, the down or d quark and the strange or s quark. The proton, for example, is a combination of two u quarks and a d quark, whereas the neutron is a combination of a u quark and two d quarks. The positively charged pi meson is a combination of a u quark and a d antiquark. Since the quark hypothesis was put forward more hadrons have been discovered and it has become necessary to add at least two more quarks, the charm or c quark and the bottom or b quark, to the catalogue of elementary particles. Nevertheless, the quark model remains a highly successful classification scheme: more than 100 hadrons are known, and they can all be described in terms of the quark model.

In spite of the success of the model in classifying hadrons, certain features ascribed to the quarks initially made the physical reality of quarks difficult to accept. The most fundamental problem is the failure to detect a free quark. The proton and the neutron are strongly bound in the atomic nucleus, yet given enough energy in a nuclear collision they can be set free. Any theory that describes the interactions of quarks, however, not only must account for their binding into hadrons but also must lead to permanent confinement.

Almost equally unsettling was the fact that in certain hadrons the quark constituents seemed to violate a fundamental principle of quantum mechanics, namely the exclusion principle of Wolfgang Pauli. The exclusion principle applies to a broad category of particles including the quarks and states that no two such particles within a small region of space can simultaneously occupy the same quan-

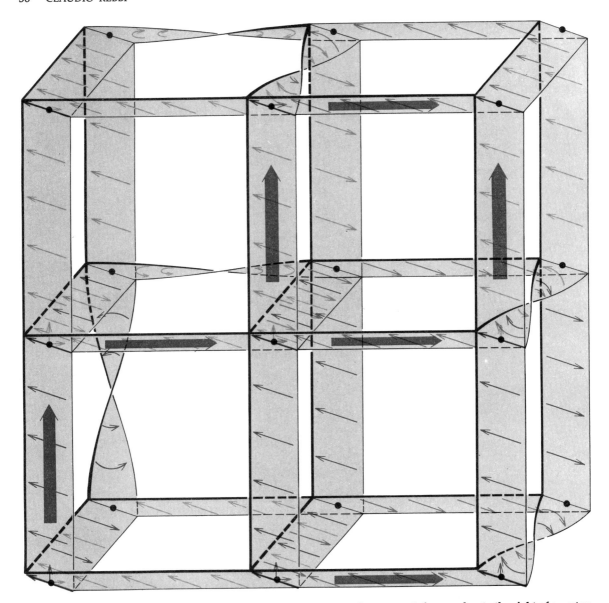

Figure 3.1 LATTICE OF POINTS helps physicists understand the field of force that gives rise to the permanent confinement of quarks. Quarks lie only at the vertexes of the lattice. The state of a particle is represented by a purple arrow, which must point in one of two directions at each vertex in the front lattice plane. To compare the orientations of two purple arrows one of them must be moved next to the other. A set of rules, represented by the bands that join neighboring vertexes, must be defined to specify the changes made in the orientation of the arrows during transport. The set of rules is a gauge field; it enables an arrow to be transported upward or to the right of a vertex, around a square array of four vertexes called a plaquette and back to its starting point. (Transported arrows are green, yellow, blue or pink.) Any set of bands that returns the transported arrow to its original orientation represents the lowest energy state of the gauge field. A set of bands that resembles a Möbius strip, however, cannot be untwisted: a transported arrow returns to its starting vertex with the opposite orientation. The red arrows represent the lines of force to which the gauge field gives rise.

tum-mechanical state. In practice the principle implies that two quarks of the same kind, say two *u* quarks, cannot form a hadron unless they have opposite spins. The spin of a quark is like the spin of the earth, except that the quark's spin is quantized: it can assume only one of two values. Hence in any group of three quarks there must be at least two with the same spin.

There are several hadrons in which three identical quarks must approach one another closely enough to bind together. The omega-minus particle is one such hadron. It was predicted by the quark model, and its subsequent discovery in 1964 by Nicholas P. Samios, Ralph P. Shutt and their collaborators at Brookhaven gave strong support to the model. On the other hand, the omega-minus was also quite puzzling because the quark model predicts it must be made up of three *s* quarks, whose close association seemed to violate the exclusion principle. For these reasons and others physicists initially preferred to regard the quark as a mathematical convenience; the question of its physical existence was temporarily set aside.

The quark model and the exclusion principle were reconciled as a result of ideas developed by Oscar W. Greenberg of the University of Maryland at College Park and, independently, by Moo-Young Han of Duke University and Yoichiro Nambu of the University of Chicago. What is needed is to assume that each kind of quark can exist in any of three states. For example, if an *s* quark in state *A* is combined with an *s* quark in state *B* and an *s* quark in state *C* to form the omega-minus particle, the exclusion principle is saved. In order to label the states of a quark physicists have whimsically taken to calling them by the names of colors (see Figure 3.2): a quark can come in the three colors red, purple and green, and an antiquark can come in the three complementary colors cyan (antired), yellow (antipurple) and magenta (antigreen). The prefix "chromo" is quantum chromodynamics refers to the color terminology.

The introduction of color had to be supplemented by another hypothesis if the successful classification of hadrons was to be maintained. Although the new color degree of freedom made possible a quark model of particles such as the omega-minus, it also led to a multiplication of other hadrons. The lambda particle, for example, is made up of a *u*, a *d* and an *s* quark; if each quark can exist in any of three colors, it would seem there should be nine lambda parti-

cles, one for each color combination, rather than the single particle that is observed. To avoid such redundancy one adds the hypothesis that the quarks in a hadron can assume only those combinations of colors that leave the hadron colorless, or white, if the rules of color addition (with ordinary light) are assumed. The three quarks in a proton or a lambda particle must include one red, one purple and one green, whereas the quark and the antiquark in a pi meson can be red and cyan (antired), purple and yellow (antipurple), or green and magenta (antigreen). Because the "total" color is always the same, in the quantum-mechanical sense that each colorless state can occur with equal probability, there is effectively only one lambda particle and only one positive pi meson.

In the late 1960's strong evidence that the quarks in hadrons are real particles instead of mere mathematical entities came from a variety of experimental results. Of notable importance was a series of experiments done at the Stanford Linear Accelerator Center (SLAC) by Jerome I. Friedman and Henry W. Kendall of the Massachusetts Institute of Technology and Richard E. Taylor of SLAC. High-energy electrons were directed against a fixed target of protons in order to probe the protons for internal structure. By examining the decay products of the collisions it was possible to show that inside the proton there are constituents with all the properties attributed to quarks.

Moreover, although no free quarks were detected, the experiments showed that within the proton the quarks are in a nearly free state of motion. This result was quite puzzling: how could forces strong enough to keep quarks permanently bound together also allow them to move about almost freely when they are at close quarters inside a proton? The three quarklike objects in the proton appeared bound to one another like the three stones tied together in a bola, the South American hunting device. The stones in the bola move freely as long as they remain within the limits of the connecting string; the string, however, keeps them from flying apart.

The new experimental evidence for quarks combined with the introduction of color gave strong impetus to the formulation of a theory of quark dynamics. Color could serve as a source for a new field called the chromoelectric field, which would give rise to a new kind of interaction among colored particles. In 1973 H. David Politzer of Cal Tech and, independently, David Gross of Princeton University

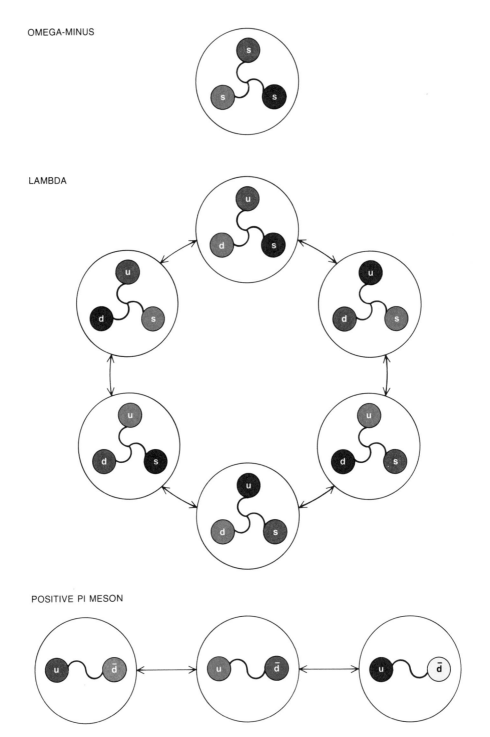

OMEGA-MINUS

LAMBDA

POSITIVE PI MESON

Figure 3.2 COLOR is a property of quarks required by the exclusion principle. The constituents of the omega-minus particle are three *s* quarks, each of which takes on one of three colors, such as red, purple and green. Color is never observed in isolation, and so the colored quarks must be combined in such a way that the omega-minus particle is colorless (or white). The lambda particle, made up of three quarks, can occupy any one of six colorless states with equal probability. The positive pi meson, made up of a quark and an antiquark, can occupy any of three colorless states with equal probability. The confinement of quarks and of color is represented by showing the quarks linked to one another by an unbreakable string.

and Frank Wilczek of the University of California at Santa Barbara realized that a dynamic interaction based on the chromoelectric field would lead to a progressively weaker force between quarks as they approached one another. The prediction explained the almost free motion of quarks inside protons that had been observed in the SLAC experiments. It was then conjectured that the same interaction could be responsible for the confinement of quarks, although at the time there were no theoretical results to support such an appealing idea. Nevertheless, once the chromoelectric field is introduced the rather ad hoc assumption that all hadrons are colorless and the observed confinement of quarks can be understood as two aspects of the same phenomenon. If one quark, say a red one, is pulled away from a hadron, both the quark and the fragment left behind are colored. If color, like electric charge, is the source of a field, there could be an attractive force between the two colored fragments. Confinement of the quarks might then result if the attraction between the two fragments is so strong that it is impossible to separate them beyond some limit.

In the early 1970's a dynamic model independent of the quark model was developed to account for certain properties of hadrons that had not been explained by quarks. According to the dynamic model, the hadron is not a point or a spherical particle but can better be understood as a string. The string can rotate or vibrate in ways prescribed by the laws of relativistic dynamics, and its end points are required to move at the speed of light. Calculations showed that the force acting along the string must be enormous: about 14 tons. The quantized vibrations of the taut string give rise to various states that could be identified with certain hadrons. A model explaining the confinement of quarks by this principle was formulated mathematically by Gabriele Veneziano of the European Organization for Nuclear Research (CERN) and interpreted as a string by Yoichiro Nambu of the University of Chicago.

It is evident that the string model of the hadron and the bola analogy can be combined. If three quarks or a quark and an antiquark are placed at the end points of the relativistic string, the tension of the string could explain the permanent binding of the particles. The string model, however, like the original concept of the quark, is itself a mere mathematical abstraction: the string is a one-dimensional object. Could it nonetheless be an approximation of some other structure that is more acceptable from a physical point of view?

In 1973 Holger B. Nielsen and Paul Olesen of the Niels Bohr Institute in Copenhagen pointed out that the string could be interpreted as a bundle of lines of force of a suitable field (see Figure 3.3). In an electromagnetic field, lines of force have the familiar patterns made by iron filings on a sheet of paper held over a magnet. The intensity of the field is proportional to the density of the lines of force. Thus when the lines of force spread out, as they do at points increasingly distant from the poles of an ordinary magnet, the intensity of the field diminishes. If the lines of force are squeezed into a tube of uniform cross section, however, the intensity of the field remains constant all along the tube (see Fig. 3.4). The force needed to separate a quark and an antiquark at opposite ends of such a tube would also remain constant no matter how far apart the two particles were placed. In order to liberate one of the quarks an infinite amount of energy would have to be supplied.

The quantum-mechanical reality of quarks and strings requires that the lines of force associated with the color interaction of quarks act quite differently from the lines of force associated with the electromagnetic interaction of electrically charged particles. Because both forces propagate in the vacuum, one might assume that any differences between them would be caused by the intrinsic nature of the forces themselves and not by the interaction of the forces with the vacuum. In classical, or Newtonian, mechanics the assumption would be sound; indeed, there can be no interaction between a field and the classical vacuum because the classical vacuum is by definition a state with no matter and no energy in it. In quantum mechanics, however, even the vacuum has a structure, which can alter the propagation of fields and forces.

T he structure of the vacuum is a consequence of the uncertainty principle of Werner Heisenberg. One version of the uncertainty principle states that for any physical event there is an uncertainty about the energy released during the event that is related to an uncertainty about the exact time of its occurrence. More precisely, the product of the uncertainty about the energy and the uncertainty about the time is not less than some numerical constant. For an event confinement to an extremely short interval there is a correspondingly large uncertainty about its energy. During any short interval, therefore, there is a substantial probability that

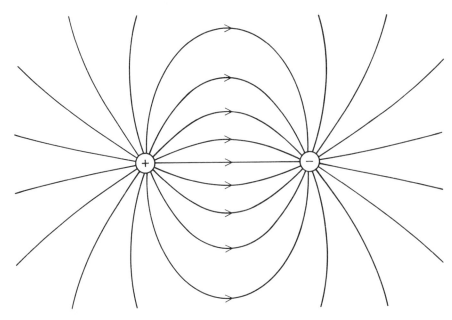

Figure 3.3 LINES OF FORCE of the electromagnetic field spread out in space. The field shown is the one generated by two particles of opposite electric charge; although the lines of force are densest in the region between the particles, they extend in other directions as well. The intensity of the field at any point (that is, the strength of the force "felt" at a given point by a unit electric charge) is proportional to the number of lines crossing a surface of unit area, orthogonal to the lines of force, that passes through the point. The electromagnetic force that is generated by a single source of electric charge diminishes as the square of the distance from the source.

the quantum-mechanical vacuum has some non-zero energy.

The energy of the vacuum can manifest itself in the spontaneous creation or annihilation of a particle and its antiparticle or in the appearance and disappearance of an electric or a chromoelectric field throughout various regions of space. Such variations of a quantum field are called fluctuations. In the electromagnetic field between two electrically charged particles, for example, the presence of quantum fluctuations implies that the interactions between the two charges are not strictly determined by the classical field predicted by Maxwell's equations. Instead the measured electromagnetic field

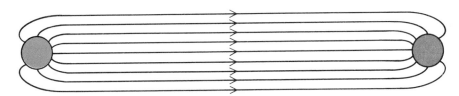

Figure 3.4 COMPRESSION of the lines of force between two particles into a thin tube of uniform cross section would make the force between the particles constant, regardless of the distance between them. A surface of unit area orthogonal to the tube would always meet the same number of lines of force, no matter where the surface was placed along the tube. Because the force that binds the particles remains constant, increasing the separation of the particles by a given increment would always require the same amount of energy, no matter how far apart they were at the outset. An infinite amount of energy would be needed to free one of the particles from the other. Such compression of the lines of force would therefore lead to the permanent confinement of the two particles. If the radius of the tube is negligible, the bundle of lines of force resembles a string.

is the average of all the fields that can be generated by the quantum fluctuations, weighted according to the probability that a given fluctuation will occur.

For most practical applications of electrodynamics the effects of the quantum fluctuations are very small. A measurement of the field between two macroscopic electrically charged objects would be consistent with the value predicted by the classical theory. In high-energy collisions of charged particles, the quantum-mechanical fluctuations become much more important and must be taken into account to calculate the electromagnetic effects. In the standard method (which for many problems is quite successful) one first calculates the properties of the field in the classical vacuum. One then builds on the result of the classical calculation by correcting it for quantum-mechanical fluctuations of a progressively higher complexity, according to what is called a perturbative expansion. In the quantum theory of electromagnetism the larger, or the more complex, a fluctuation, the less the probability that it will take place. Hence almost all the corrections to the classical electromagnetic field that must be made in the quantum-mechanical calculations are the outcome of small fluctuations.

One might suppose the properties of the chromoelectric field between two quarks could be deduced in a closely analogous way. It would seem, at least in principle, that a perturbative expansion could give the strength of the field at any point to any degree of accuracy needed. It turns out, however, that the method of perturbative expansion works only if the field calculated for the classical vacuum is the dominant effect. In other words, the method works only if the corrections that must be made to take account of the fluctuations are small and become smaller still as fluctuations of increasing size are considered. For quantum-mechanical phenomena that depend primarily on the effects of large fluctuations the perturbative expansion does not converge, that is, the series of calculations does not approach a constant, finite value. Such phenomena are said to be nonperturbative. The confinement of the lines of force of the chromoelectric field between two quarks, and hence the permanent binding of the two quarks, is a nonperturbative phenomenon.

How, then, can confinement be demonstrated? Workers in theoretical physics recognized that a new approach had to be devised, in which large fluctuations of the quantum-mechanical field are considered at the outset of the calculations. The lattice method suggested by Wilson is such an approach.

The lattice is generally a cubic one and can be thought of as the edges and vertexes of a collection of densely stacked cubes. The lattice extends in time as well as in space, that is, each point on the lattice designates both a position in space and a moment in time. To visualize the lattice one can think of an array of cubes in which two of the axes are labeled with spatial coordinates and the third axis is labeled with temporal coordinates; the full lattice has three spatial dimensions as well as the time axis, and so it is a four-dimensional structure. Between any two neighboring vertexes of the lattice there is a link, which can be pictured as a line connecting the two vertexes. A small square bounded by four links is called a plaquette. In Wilson's formulation the vertexes, links and plaquettes of the lattice are all that is left of ordinary physical space and time.

Links and plaquettes on the lattice must be regarded as entities at a different level of abstraction from the vertexes, or lattice points. Although links and plaquettes are defined by lattice points, there are no additional lattice points along a link or within a plaquette. In other words, space and time on the lattice are quite unlike ordinary space and time, which always include an infinite number of points between any two given points.

Wilson's introduction of a space-time lattice is not meant to imply that physical processes really take place on a lattice. Space-time, according to all current evidence, is continuous. Instead the lattice represents what theoretical physicists call a regularization, a temporary artifact for making calculations that would otherwise be impossible. Applied to the problem of confinement, the strategy is as follows. All particles are defined only at the vertexes of the lattice, and the strength of the field is defined only along the links of the lattice. (Actually what is defined at each vertex is the probability that a particle will be found there. The probability of finding a particle between two adjacent vertexes is not defined.) When no particles are present, the symmetry of the fluctuations on the lattice dramatically simplifies the calculation of the average electric or chromoelectric field generated by strong fluctuations. The fields, which are vector quantities and therefore have both magnitude and orientation, are just as likely to point in one direction along a link as they are to point in the opposite one. Hence in the vacuum state, with no particles, the mean value of the electric or the chromoelectric field throughout the lattice is zero.

A similar although slightly more elaborate calculation can be done for large fluctuations of the field when a single particle and its corresponding antiparticle are defined on the lattice. On the average the fluctuations of the field again cancel except along the links of the lattice that make up the shortest path between the particle and the antiparticle. The results do not depend on the kind of field defined on the lattice; the particles can be a quark and its antiquark or an electron and a positron. Thus confinement is a natural outcome of defining the field on the lattice.

The next step in the strategy is to remove the lattice and regain ordinary, continuous space-time. The lattice spacing is made progressively smaller so that the vertexes of the lattice become closer and denser in space-time. If the reduction in the lattice spacing proceeds to the mathematical limit, continuous space-time is recovered. At the time the procedure gives the average field after all the quantum-mechanical fluctuations have been taken into account.

What is gained by introducing the lattice? The strong fluctuations that must be responsible for squeezing together the lines of the color force are considered from the outset along each lattice link. On the other hand, there is a price that must be paid: as long as the lattice is relatively coarse the quantum fluctuations give rise to the confinement of the electromagnetic field as well as the chromo-electric field.

It is well known that electric charges, unlike color charges, do exist in isolation. Since the lattice method predicts the confinement of electric charge, one must be skeptical about the lattice approach unless it can be shown that as the mesh of the lattice is made finer the electromagnetic field begins to act in accordance with its well-established properties. In other words, what one would like to show is that at some stage in the shrinking of the lattice the electromagnetic lines of force break out of their confinement to a line of lattice links, whereas the chromoelectric lines of force remain squeezed together all the way to the limit of continuous space-time. It is not a straightforward matter to demonstrate that things happen in precisely this way, but in the past few years the demonstration has been achieved by making elaborate numerical calculations with the aid of high-speed computers.

Theoretical physicists express the fundamental difference between the electromagnetic field and the chromoelectric field by saying that QED is an Abelian gauge theory whereas QCD is a non-Abelian gauge theory. The terms refer to the Norwegian mathematician Niels Henrik Abel. The distinction between Abelian and non-Abelian is drawn from the mathematical theory of groups, which describes the symmetries inherent in a sequence of operations, such as a sequence of rotations. If the operations that are members of a group can be carried out in any sequence with the same final result, the group is Abelian. For example, the group of rotations about a single axis is Abelian, because such operations have the same effect regardless of their sequence. On the other hand, if the sequence in which two or more operations are carried out does affect the final outcome, the group of operations is a non-Abelian one. The rotations of a cube about its three axes form a non-Abelian group: when the cube is turned about a vertical axis and a horizontal one, the result depends on which operation is done first.

In order to understand how ideas from group theory apply to QCD and QED, one must understand the concept of a gauge field. The concept can best be illustrated for isolated points of space and time arranged in a lattice. Particles can rest on any of the lattice vertexes or hop from one vertex to another; as the particles move through the space-time lattice they can change their state, where a state is defined by quantities that can vary over a certain range of values. I shall make the simplifying assumption that the state of a particle is described by just one variable, which can take on exactly two values. For example, at each vertex of the lattice there might be a variable whose value is either +1 or −1, indicating the sign of the electric charge. The two possible values can be represented by an arrow that points either up or down.

In describing how interactions propagate in space and time it is essential to be able to compare the values of variables at neighboring points. To compare the length of two objects in separate regions of space one needs a measuring stick, or gauge, that can be moved next to one object, marked and then moved to the second object, where the comparison is made. Similarly, on the lattice one must be able to compare the orientation of the arrows at neighboring vertexes. At first such a comparison seems trivial. Suppose two vertexes of the lattice are represented on a sheet of paper and at each vertex there is an arrow directed upward. Is it not obvious that the two arrows point in the same direction? The question, however, presupposes the existence of the sheet of paper on which the lattice is drawn. The paper acts as an intermediary that allows the two orientations to be compared. In a sense, one transports the arrow from one vertex to the other with the eye, and one concludes that the two arrows had the same orientation before the transport because after the transport they match.

Suppose the sheet of paper between the two arrows had been given a half twist (see Figure 3.5). An upward-pointing arrow transported along the twisted paper would point downward when it reached the second vertex. Because the direction of an arrow is defined only at the individual, isolated vertexes of the lattice, there is no way to decide which of the two methods of transport is correct. Indeed, without the sheet of paper or some similar

assumption about the effects of transport no comparison of directions at different vertexes is possible.

In field theory a gauge is any standard of measurement, analogous to the distance between two marks on a metal bar or the direction of an arrow with respect to a dial, that can change under the influence of a field as the gauge is moved about in space and time. A field that can effect such changes is called a gauge field, and it specifies explicitly the assumptions that must be made about the transport of the gauge. In my simple example the gauge field is a set of rules for transporting the arrows along the links of the lattice from one vertex to the next. One can think of the gauge field as a ribbon that connects neighboring lattice sites and thereby allows the arrows at different points to be transported and compared.

It is important to note that the arrow representing

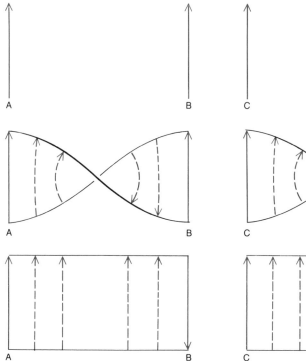
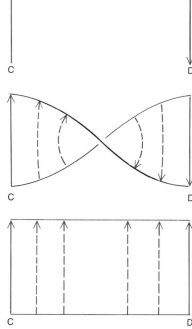

Figure 3.5 DIRECTIONS OF TWO ARROWS defined only at the vertexes of a lattice cannot be compared unless there is some way to transport the arrows. In the top diagram the arrows at A and B seem to point in the same direction, whereas the arrows at C and D seem to point in opposite directions. The comparison presupposes the existence of the paper on which the diagram is drawn, which acts as an intermediary along which one arrow can be moved next to the other. One might also assume that the points in each pair are connected by a half-twisted ribbon of paper (middle); the comparison then gives the contrary result. Without a gauge field to supply a rule of transport from point to point the orientations of the arrows at adjacent points cannot be compared. The bottom diagram shows that the result of a comparison need not change when the direction of an arrow is reversed at any point; the comparison remains valid if the ribbon is correspondingly twisted or untwisted.

the state at any lattice vertex can be reversed, providing the ribbon representing the gauge field is correspondingly twisted or untwisted. If the arrows and the ribbons are adjusted in coordination, the result of any comparison between state variables does not change, nor does the physical information represented by the system. The possibility of reversing the arrows without modifying the physical information is called local gauge invariance.

Because of local gauge invariance one might think a gauge field is nothing but an unnecessary complication. What is the point of introducing the twisted ribbon if it can then be untwisted by a local gauge transformation? The objection is valid as long as one considers only pairs of lattice points. The importance of the local gauge invariance becomes apparent, however, when one considers a plaquette on the lattice: a set of four points in a square, connected along the lattice links by four ribbons that constitute the gauge field. Suppose one of the ribbons is twisted and the other three are not (see Figure 3.6). The single twisted ribbon can be untwisted (and one of the arrows can be reversed), but not without introducing a half twist in one of the other three ribbons. No matter which arrows are reversed and no matter how many local gauge transformations are carried out, an odd number of the four ribbons must be twisted. A plaquette whose ribbons cannot all be untwisted is called a frustrated plaquette.

The frustration of the plaquette manifests itself in another way. An upward-pointing arrow that is transported along the ribbons around the frustrated plaquette will return to its starting vertex pointing down. Thus on the lattice the frustrated plaquettes are those that generate a mismatch in orientation when an arrow is transported all the way around.

It is not difficult to apply the idea of a frustrated plaquette to ordinary space-time. In physical space-time the ribbons along which the arrows move cannot be visualized directly; just as the sheets of paper that define the gauge field on the lattice are not themselves part of the lattice, so a gauge field defined in ordinary space-time is not itself part of space-time. Mathematically the higher-dimensional abstract space that specifies the rotations of the arrow is called a connection in a fiber bundle [see "Fiber Bundles and Quantum Theory," by Herbert J. Bernstein and Anthony V. Phillips; SCIENTIFIC AMERICAN, July, 1981]. Nevertheless, it is possible without complete mathematical understanding to imagine that an arrow moved about along some closed path in space-time could return to its starting point with its direction changed.

The idea of a frustrated plaquette has an important physical interpretation. In any gauge field the energy of the field resides precisely in the plaquettes that are frustrated. The plaquettes that are frustrated, in which all the ribbons can be untwisted and all the arrows can be oriented in the same direction, are associated with the vacuum state of a physical system, the configuration with no energy. Frustrated plaquettes are the sites of the fluctuations in a quantum-mechanical field.

The idea of a gauge field can readily be general-

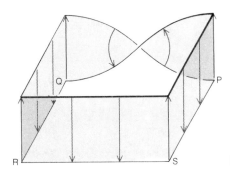

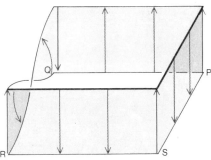

Figure 3.6 FRUSTRATED PLAQUETTE is given a unit twist by the four gauge fields that make up its sides. The gauge fields, represented by ribbons, can be twisted or untwisted locally to compensate for the reversal of an arrow at any point, but the overall twist in the four ribbons of the plaquette cannot be removed. The twist between P and Q (left) can be removed by reversing the direction of the arrow at Q. The effect, however, is merely to transfer the twist to the part of the ribbon between Q and R (right). No matter which arrows are reversed, or in other words no matter how many local gauge transformations are performed, an odd number of the four ribbons must stay twisted. Hence the plaquette remains frustrated in spite of local gauge invariance.

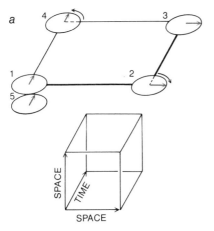

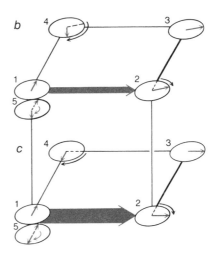

Figure 3.7 IN THIS EXAMPLE, THE FRUSTRATION OF PLAQUETTES is associated with a single component of the electric field, the spatial component in the left-right direction. As a dial is transported counterclockwise around a plaquette that includes the left-right spatial component and the time dimension, the gauge field causes an arrow on the dial to rotate. If the arrow returns to its starting position after it makes a full circuit of the plaquette, the strength of the field is zero in the spatial direction and there is no field line (*a*). As the phase angle increases, the strength of the field along the spatial direction increases (*b-c*), which is represented by the increasing thickness of the field lines (*colored arrows*) along the spatial coordinates.

ized to more complex situations. The arrows, for example, can be allowed to form an arbitrary angle with respect to a fixed direction rather than being constrained to point up or down. The gauge field now specifies the angle by which the arrow rotates when it is transported along a link (see Figure 3.7). Consequently the frustration of a plaquette can take on a continuous range of values, measured by the angular difference in the direction of the arrow after it is transported around the entire plaquette. The degree of frustration of a plaquette, expressed in suitable mathematical units, is called its action. The action of the entire system is the sum of the actions of all the individual plaquettes.

It turns out the electric field is a gauge field that specifies the continuous rotations of an arrow about a single axis, although this fact is not apparent in any but the most unified and sophisticated formulations of the concept of the electric field. It is more common to think of the electric field as a set of vectors, with one vector at each point in space giving the magnitude and direction of the field at that point. Understood according to the more comprehensive framework of the gauge theory, however, the magnitude of each vector is directly proportional to the frustration of an associated plaquette.

The rotating arrow of the gauge field associated with electromagnetism actually represents a complex number (one with both a real part and an imaginary part), which changes in value as the arrow is carried around the plaquette. (The transported arrow and the complex number must not be confused with the vector that represents the electric field itself at each link on the lattice.) The plaquette is a loop in space-time, and so it is necessary to imagine the transported arrow as moving forward and backward in time as well as in space. The arrow is carried along a certain axis in space, say the positive x axis, then forward in time, then back to its starting point on the x axis and finally backward to its starting point in time. The amount by which the direction of the transported arrow changes as a result of the transport is called the phase angle. According to gauge theory, the magnitude of the x component of the vector that ordinarily gives the strength of the electric field at a point is a measure of the phase angle generated around a space-time loop that begins in the x direction. The quantum-mechanical fluctuations at a point in the electric field can therefore be thought of as fluctuations in the amount of rotation an arrow would undergo if the arrow were transported around a plaquette extended along one spatial dimension and along the temporal dimension (see Figure 3.7).

It is difficult to suppress a feeling of unreality when one is asked to regard the electric field—an entity that can become quite tangible when one gets a shock—as the abstract space of phase rotations. Nevertheless, as one goes more deeply into the study of the physical world, tangible facts and mathematical concepts become intertwined. The abstract idea of phase space at last makes contact with the distinction between QED and QCD, that is, with the distinction between an Abelian gauge theory and a non-Abelian one. The fundamental group operations in QED are the rotations of phase angles. The result of two successive phase rotations, say one of 30 degrees and one of 50 degrees, does not depend on the sequence in which they are done. The net result in either case is a rotation by 80 degrees. Such an outcome is characteristic of an Abelian group and QED is therefore an Abelian gauge theory.

If the state variables defined at the vertexes of a lattice are the colors of quarks, the transported gauge is made up of three arrows instead of just one, and the arrows can vary in length as well as in angle (see Figure 3.8). Each arrow represents one of three color charges; the color charges together determine the color of the quark. As the quark is transported along a link of the lattice it can change its color from red to green; hence a red quark can exchange colors with a green quark, and then the newly green quark can exchange colors with a purple quark. The final result of the exchanges, however, depends on the sequence of events. Since an outcome that depends on sequence is characteristic of a non-Abelian group, QCD is a non-Abelian gauge theory (see Figure 3.9).

The non-Abelian nature of QCD introduces an extra degree of freedom into the fluctuations of the chromoelectric field. Moreover, the existence of three kinds of color also implies that the plaquettes of the chromoelectric field can be frustrated in many more ways than the plaquettes of the electromagnetic field can. It is likely that the extra degree of freedom and the additional disorder of the non-Abelian theory are responsible for the confinement of quarks within hadrons.

I have already stated that confinement on a coarse lattice can be demonstrated by a relatively simple calculation for large fluctuations of the quantum-mechanical field (see Figure 3.10). For smaller fluctuations on a finer lattice, however, the calculation of the field becomes much more difficult. The value of any physical quantity measured in an experiment is the quantum-expectation value, which is a weighted average of all possible values the quantity can have. The measured electric or chromoelectric field is the average of all the possible configurations of the field to which fluctuations can give rise. Not all configurations contribute equally to the average, and so each configuration must be weighted, or multiplied by some factor based on the probability of the configuration. In principle, therefore, the quantum-expectation value of a physical quantity defined on the lattice can be calculated in two steps. First the value is calculated for each configuration of the field fluctuations and multiplied by the weight factor for that configuration. Then the products are added and the result is divided by the sum of the weight factors.

Even for a lattice defined over a small volume of space-time, however, the number of possible configurations is so large that a complete summation is out of the question. For the simplest-possible gauge theory, in which the gauge field between two lattice vertexes can be either twisted or untwisted, the number of configurations on a lattice extended for 10 sites in each direction is $2^{40,000}$, or more than $10^{12,041}$.

To calculate the quantum-expectation value of the field on the lattice one must therefore resort to techniques of statistical sampling. The techniques are analogous to the ones employed in opinion polling. One cannot ask the opinion of every person in the U.S. in order to determine how an issue is perceived by various groups in the population, and so a sample is selected. The result should reflect the actual opinion of the groups if the probability of selecting a respondent from a given group matches the portion, or weight, of the group in the population as a whole.

In a similar way the configurations of the field fluctuations are sampled by a computer program. The computer generates a large number of configurations (but many fewer than $10^{12,041}$), and the probability that a particular configuration is generated is set equal to the quantum-mechanical weight factor for that configuration. The sample of configurations tends to give the same average value for the quantum field as the total population of configurations does.

The importance of a configuration in calculating the quantum-expectation value is determined by its action, which is generally designated S; the weight factor is therefore given by a mathematical function

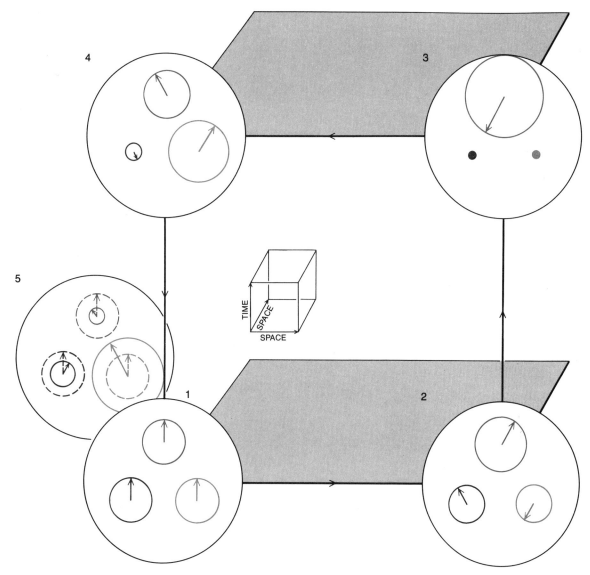

Figure 3.8 CHROMOELECTRIC FIELD is a gauge field similar in principle to the electromagnetic field but more complicated mathematically. At every point on a lattice there are three arrows instead of one; they correspond to the color charges of a quark. Moreover, the color gauge field affects not only the direction of each arrow but also its length. The lengths of the arrows are not independent of one another: the square root of the sum of the squares of the lengths must equal 1. The strength of the chromoelectric field along each link of the lattice is dependent on the phase angles and the change in the configuration of the three arrows after a complete circuit of a plaquette.

of the action S. An elegant formula for the weight factor in quantum-mechanical systems was introduced by Richard P. Feynman of Cal Tech. First the total action of the configuration is divided by a constant, g^2, then the exponential of this quantity is determined; in other words, the number e (equal to approximately 2.7) is raised to the power S/g^2. The weight factor is inversely proportional to the result. Thus the weight factor is proportional to $1/\exp(S/g^2)$.

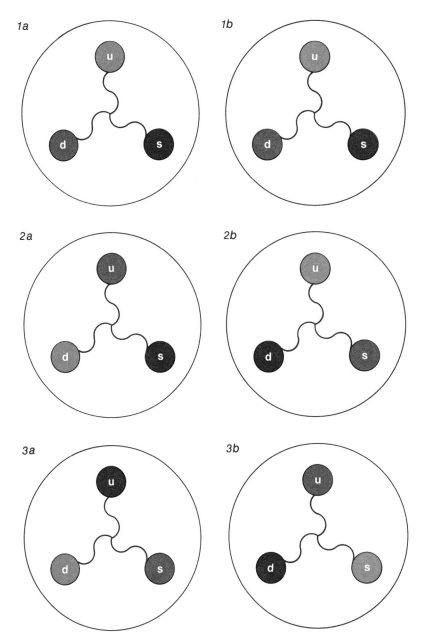

Figure 3.9 GROUP OF COLOR TRANSFORMATIONS in the chromoelectric field is classified mathematically as a non-Abelian one. A non-Abelian group is an abstract set of operations (together with the objects on which the operations are carried out) in which the final outcome of several operations is dependent on the sequence in which they are done. For the chromoelectric gauge theory the operations are interchanges of colors among quarks. In the illustration two interchanges of color are applied to a given quark configuration (1a, 1b). At the left an exchange of red and green (2a) is followed by an exchange of purple and red (3a). At the right the order of the two exchanges is reversed (2b, 3b). The final configurations of the quarks are not the same.

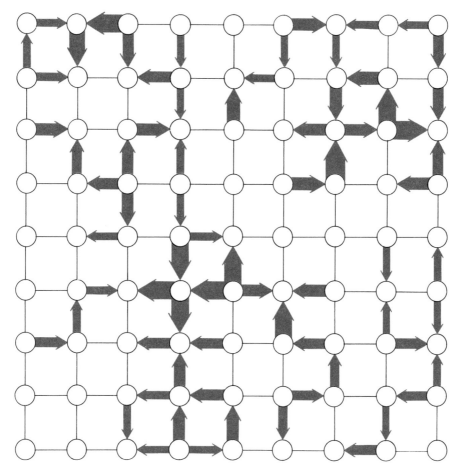

Figure 3.10 FLUCTUATIONS of the lines of force on a lattice are characteristic of the quantum-mechanical vacuum and are a consequence of the uncertainty principle. During any short interval there is a correspondingly large uncertainty about the energy of the system, which manifests itself in spontaneous fluctuations of the quantum field. The fluctuations cancel one another in the absence of charged particles, and so a measurement of, say, the electric field in the vacuum would give the value zero because it would determine only the average of all the fluctuations. Nevertheless, the fluctuations of the lines of force in the vacuum must be carefully considered in calculating the field of force between two charged particles.

The formula for the weight factor implies that the higher the action of a configuration is, the less the configuration is weighted in calculating the average quantum field. Because the action S is not itself exponentiated but instead is divided by g^2, changing the value of g can have a significant effect on the weight factor of a given configuration. If g is large, S/g^2 is small and $1/\exp(S/g^2)$ is larger than it would be if g were small. The quantity g is called the coupling constant; hence when the coupling constant is large, the weight factor of a quantum-mechanical fluctuation with large action is higher than it is when the coupling constant is small.

The idea of a coupling constant may be familiar as a measure of the intrinsic strength of a force. In electromagnetic theory the coupling constant is an important physical quantity with a value of about 1/137. From the rather abstract perspective of the lattice gauge theory, however, g is to be understood as a quantity that takes on a fixed value for a given lattice spacing but can vary with the spacing

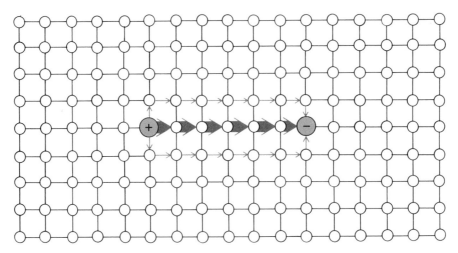

Figure 3.11 CONFINEMENT OF A FIELD, whether the field is the electric or the chromoelectric one, can be demonstrated by fairly simple calculations when the field is defined on a coarse lattice and the fluctuations of the lines of force are strong. The strong fluctuations cancel one another everywhere on the lattice except in the region between two charged particles. The lines of force are compressed into a thin tube, and so the two particles cannot be moved indefinitely far apart.

(see Figure 3.11). Once confinement is demonstrated on a coarse lattice the mesh of the lattice is made finer by carefully decreasing the value of the coupling constant. The process by which the lattice is made progressively finer until continuous space-time is recovered is called renormalization (see Figure 3.12).

Strictly speaking, it is not yet possible to renormalize the lattice for fluctuations of the chromoelectric field, or in other words to shrink the lattice spacing

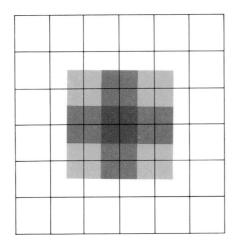
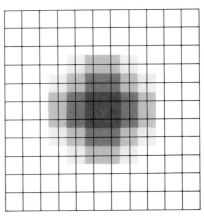

Figure 3.12 SHRINKING A LATTICE on which particles and fields have been defined must be carried out in order to approximate the effects of the field in ordinary, continuous space and time. As the lattice spacing is reduced, however, the size of a composite particle such as a proton must be preserved, that is, the probability that a quark will be found at a given lattice vertex within the proton must be spread over an increasing number of vertexes. The illustration at the right shows the result of shrinking the lattice at the left to half its previous size. The color density represents the relative probability that a quark will be found at a given vertex. The shrinking can be accomplished by systematically changing the value of the coupling constant.

to zero. Nevertheless, it is possible to make the lattice spacing smaller while still keeping it greater than zero and to search for indications that the lines of force do or do not remain confined on finer lattices. As the lattice spacing is reduced all physical objects should remain the same size. On a coarse lattice the probability distribution for a proton might be defined as nonzero across only three lattice spacings. When the lattice spacing is reduced to half its original value, the probability distribution must stretch over six lattice spacings. Since reducing the value of g lowers the probability of lattice configurations that have a large action, the decrease in g has the effect of "zeroing in" for a closer look at the fluctuations of the field, at the lines of force and at the particles defined on the lattice. Thus reducing the value of g makes the proton look larger on the lattice.

The numerical investigation of the properties of a gauge field on the lattice proceeds by limiting the lattice to a large but finite volume. The number of lattice vertexes, links, plaquettes and state variables is therefore finite, although it may be larger than 100,000. Initial values of all the variables are stored in the memory of a large computer. By randomly varying the elements in the starting configuration according to a suitable algorithm the computer generates a sample of as many as 100,000 configurations. Finally it calculates the average quantum-mechanical effects to which the configurations in its sample give rise.

B ecause of the element of randomness in the calculation, the method is called a Monte Carlo simulation. Before my own work and that of my colleagues on quark confinement the Monte Carlo method had been applied with considerable success to the analysis of the properties of thermodynamic systems, and Wilson had emphasized its suitability for the analysis of lattice gauge theories in quantum mechanics. In 1979, working at Brookhaven, Michael J. Creutz, Laurence A. Jacobs and I first applied Monte Carlo simulation to the study of Abelian gauge theories. We wanted to test whether or not the confinement of particles on the lattice observed at large values of the coupling constant in QED disappears as it should when the continuum limit is approached. The results were spectacular; they showed clearly that at some stage in the reduction of the coupling constant the lines of force on the lattice suddenly undergo a transition. The electric field, which for large values of g is confined to the lattice links between two electric charges, suddenly spreads out all around the charges. (see Figure 3.13).

It is useful to compare the sudden deconfinement

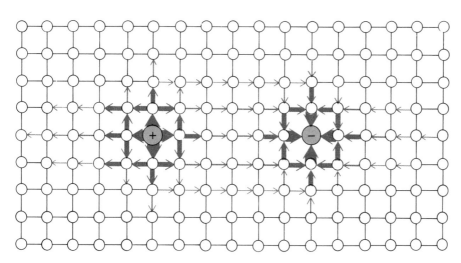

Figure 3.13 CALCULATION OF THE ELECTRIC FIELD generated by two oppositely charged particles defined on a lattice. The illustration shows the outcome of thousands of calculations for progressively finer lattice spacings. As the spacing is reduced the lines of force that are confined on the coarse lattice (see Figure 3.11) suddenly undergo a phase transition and break out in all directions. Similar calculations done by Cruetz have demonstrated that as the lattice spacing is reduced for the chromoelectric field the lines of force remain confined, just as they are on the coarse lattice. Thus the confinement of quarks can be derived from the theory of the chromoelectric field.

of the electric field with the sudden change in properties observed when a solid changes to the liquid phase. What Creutz, Jacobs and I observed in our numerical calculations was a phase transition. Everything happened in the computer model as if we had a four-dimensional crystal that we could heat or cool by changing the value of g. At a certain point the crystal underwent a phase transition; on one side of the transition electric charges are confined, whereas on the other side they are not. Our results were later confirmed independently by an extensive numerical simulation done at the European Organization for Nuclear Research (CERN) by Benny Lautrup of the Niels Bohr Institute and Michael Nauenberg of the University of California at Santa Barbara and by other investigators.

Soon after we got our results for the Abelian lattice gauge theory Cruetz extended the Monte Carlo simulation to the non-Abelian model. His results were just as spectacular and physically more interesting; with QED the correct answer was known before hand and the simulation was a test of the Monte Carlo method rather than of QED, but the outcome for a non-Abelian system was entirely unknown. Creutz's simulation showed that, contrary to the Abelian case, the non-Abelian lattice gauge theory undergoes no phase transition as the value of the coupling constant is gradually lowered. Thus what had long been sought is now finally obtained: a demonstration, albeit by numerical methods, that quantum chromodynamics leads to the confinement of quarks.

Monte Carlo simulation makes it possible to explore the predictions of QCD for many physical processes. For example, after Cruetz showed that there is no deconfining phase transition in QCD he was also able to estimate the force holding the quarks together. The result is in excellent agreement with the prediction of the string model, which makes no pretense of describing the dynamics of quarks and hadrons in their full generality. Creutz's value for the string tension was confirmed independently by Wilson, by Gyan Bhanot of CERN, by me and by several others.

Another prediction of Monte Carlo simulation is the surprising result that at extremely high temperature quarks become deconfined and can move about freely. The temperature must be about two trillion degrees Celsius, much too high to be created in the laboratory but perhaps not too high for free quarks to be found in the interior of hot stars. The prediction also suggests that free quarks existed shortly after the big bang.

The current frontier in the understanding of quark interactions is the numerical simulation of changes in the chromoelectric field as the quarks move about. All the investigations described so far evaluate the chromoelectric field by assuming that the quarks are stationary. It is possible to employ the Monte Carlo method to simulate the quantum mechanics of moving quarks in a gauge field, but the amount of computation required puts the simulation almost out of reach, even for the most powerful computers. Herbert W. Hamber of the Institute for Advanced Study, Enzo Marinari and Georgio Parisi of the University of Rome and I, and independently Donald H. Weingarten of Indiana University, have recently proposed an approximation scheme whereby the computations become feasible. The approximation has been applied to a calculation of the masses of several hadrons, and the theoretical results are in good agreement with experimental values.

Lattice gauge theory has at last brought QCD to the stage where one can calculate its predictions and compare them with experiment. It is an elegant theory, based on relatively simple mathematical concepts yet rich in consequences. For the moment all its verifiable predictions are passing the test of experiment. What remains to be done is to learn how its consequences can be proved by logical deduction. Although a numerical approximation cannot satisfy the need for logical demonstration, it does augur well for such a demonstration in the future. In other branches of theoretical physics there have been many instances in which a definite result, obtained numerically at first, was soon proved by analytic methods. Several of my colleagues and I are confident that QCD will soon pass this test as well.

Postscript: Section I, Ideas

· · ·

Chris Quigg

The standard model of elementary particle physics I described in Chapter 1 continues to account for all experimental observations, to suggest fruitful new observations and to serve as a point of departure for new explorations. Many of the hopes and dreams expressed in Chapter 2 by Sheldon Lee Glashow are now the firmly established foundations of our understanding. Over the past several years, remarkable progress has been made in applying quantum chromodynamics (QCD), the theory of strong interactions, to experiments and in extracting detailed, precise information about the high-energy interactions of quarks and gluons from observations of the products of proton-antiproton collisions. Still more incisive experimental studies are anticipated from the 2 TeV Tevatron Collider at Fermilab.

Steady but less dramatic progress has been made in deducing the consequences of QCD in the low-energy (or long distance) regime where the strong interactions are strong, the realm of hadron structure. In order to deal with the existence and properties of the hadrons themselves, it is necessary to devise a new computational strategy that does not break down when the interaction becomes strong. The most promising approach has been the crystal lattice formulation of QCD described in Chapter 3 by Claudio Rebbi, in which space-time is viewed as a discrete, rather than a continuous, structure. This makes it possible to exploit many of the techniques developed earlier in statistical mechanics for the study of spin systems such as magnetic materials.

One of the most valuable implementations of this strategy for studying hadron structure uses computer simulations in which different gluon field configurations are explored by random sampling Monte Carlo techniques. These techniques make heavy demands on computer time and have, in turn, spurred the development of new computer architectures. Further, these calculations have yielded suggestive evidence that quarks and gluons are indeed permanently confined in QCD. The ultimate hope is to compute the spectrum and properties of hadrons from first principles.

The search for a new state of matter, the quark-gluon plasma, which might form when hadronic matter is compressed and heated to extreme temperatures, is beginning in earnest, using high-energy heavy-ion collisions. Understanding what questions to ask of this new kind of experiment is an important priority for the immediate future.

The success of the standard model challenges us to examine its self-consistency in order to create a more complete "theory of the world." To approach this challenge, we look to the standard model itself for guidance. The "Open Questions" cited in Chapter 1 are among the larger issues that guide this speculation and experimentation. The shortcomings of the standard model may also be expressed in somewhat different form. The sharpest issue posed by the standard model is our incomplete understanding of the scalar, or Higgs, part of the electroweak theory, that is, the mechanism that conceals the electroweak gauge symmetry. A second concern is the meaning of quark-lepton generations. The idea that quarks and leptons should be grouped in generations seems necessary for the internal consistency of the electroweak theory, but we do not know why this should be so. Third, we may even dare to question the origin of the gauge symmetries themselves. Such questions—and this is but a partial list—stimulated by the standard model itself, continue to drive our desire to find ever simpler, more general descriptions of nature.

One possible solution to the Higgs problem introduces a completely new set of elementary particles with spins differing by one-half unit from those of the known quarks, leptons and gauge bosons. These postulated new particles are the consequences of a conjectured "supersymmetry" that relates particles of integer and half-integer spin. Supersymmetry cannot be an exact symmetry of the world in which we live, for that would imply, for example, a spinless counterpart of the electron with the same mass that would have been observed long ago. Supersymmetry would stabilize the Higgs boson mass below 1 TeV and the supersymmetric partners of the known particles would have masses less than 1 TeV. Part of the appeal of supersymmetry is that its local form leads to quantum theories that have Einstein's gravitation as their classical limit. This raises the possibility of obtaining a quantum theory of gravity and of incorporating gravity into a unified description of all the fundamental interactions.

Extensive exploratory searches for superpartners at accelerators have been negative so far. They have restricted specific models of sypersymmetry, but have not ruled out supersymmetry as a means of resolving the problem of electroweak symmetry breaking. Negative results become decisive only after a thorough exploration of the 1 TeV scale, one of the principal goals of the Superconducting Super Collider. The machines now coming into operation will only extend the search to a few tenths of a TeV, where the first superpartners could be found.

A second possible solution to the Higgs problem is based on the idea that the Higgs boson is not an elementary particle, but a composite made out of elementary constituents analogous to the quarks and leptons. Resembling the usual quarks and leptons in certain respects, these new constituents would have a new type of ultrastrong interaction—often called "technicolor"—that would confine them within about 10^{-17} centimeters. The new phenomena would include a rich spectrum of bound states, akin to the spectrum of known hadrons. Although there is no evidence yet for any of these new particles, the appeal of the technicolor scenario is that it goes beyond the Higgs mechanism of the standard model in much the same way as the Bardeen-Cooper-Schrieffer theory of superconductivity gives a microscopic interpretation of earlier phenomenological descriptions of the superconducting state.

Other theoretical inventions deal with the meaning of generations and the origin of gauge symmetries. These, too, imply new kinds of matter and force particles. A particularly audacious initiative is the search for a "superstring" theory, described in Chapter 11 by Michael Green, which would solve all the outstanding problems while unifying gravitation and the other forces of nature. The observation of supersymmetric partners at or below 1 TeV would provide powerful encouragement for superstrings, for which supersymmetry is an essential foundation.

Theoretical speculation of this sort is helpful and often indispensable, but without experimentation theory is sterile. It is important to follow up hints given by theory, for one of the great lessons of the history of physics in the 20th century is that respect for great principles pays rich dividends. However, it is also important that the prepared mind—and the prepared apparatus!—go where no one has gone before and *explore*. The discoveries recounted in Chapters 6 and 8 by Martin Perl, William Kirk and Leon Lederman did not result from treasure hunts inspired by theoretical models. An important motivation for the Super Collider is precisely this: to take a significant step beyond the present confines and to learn what is there. The progress of our science demands it.

SECTION

II

TOOLS

. . .

The Next Generation of Particle Accelerators

The smallest of objects can be perceived only with the largest of instruments. Penetrating deeper into the structure of matter may require accelerators built with multinational sponsorship.

. . .

Robert R. Wilson
January, 1980

For some 50 years the effort to understand the ultimate structure of matter has proceeded almost entirely through a single experimental technique. A particle of matter is brought to high speed and made to strike another particle. From an examination of the debris released in the aftermath of the collision, information is gained about the nature of the particles and about the forces that act between them. To carry out a program of such experiments it is necessary to have a source of energetic particles. Cosmic rays provide a natural source, but the flux of particles is diffuse and is beyond the control of the experimenter. A more practical source is a particle accelerator, the device for increasing the speed of a particle and hence also its energy.

One of the first particle accelerators, built by Ernest O. Lawrence in 1928, was made of laboratory glassware a few inches in diameter. Most of the accelerators in service today are lineal descendants of Lawrence's device, but they have grown enormously in size and complexity. The largest extend over many square kilometers and indeed are so large that the availability of real estate has become a significant consideration in their design. The particle accelerator is no longer an instrument installed in a laboratory; instead the laboratory is assembled around the accelerator. Building such a machine costs hundreds of millions of dollars; operating it requires a staff of about 1,000 people and dozens of digital computers.

A new generation of particle accelerators is now in prospect. The first few are just coming into operation; several more are under construction; others are still being planned, and their characteristics are not yet fixed. For both the physicist and the layman the principal interest inspired by these new machines is in the results of the experiments they will make possible, but the accelerators themselves also merit notice. In the physics of elementary particles the highest available energy represents a frontier marking one of the boundaries of experimentally verifiable knowledge. Several of the new accelerators will be capable of attaining higher energies than any existing machine, and so they will push the frontier into unexplored territory. In order to reach those energies the accelerators will of necessity be larger, more complicated and more expensive than their predecessors.

Largely because of the cost, the construction of an accelerator today requires the resolution not only of technical problems but also of political, economic and managerial ones. Money for scientific research

Figure 4.1 TANDEM ACCELERATORS occupy a single tunnel at the Fermi National Accelerator Laboratory. The blue objects in the upper tier are the beam-bending magnets of the original "main ring" proton synchrotron, which began operating in 1972 and has reached a peak energy of 500 billion electron volts (GeV). Lower tier of red magnets is the first segment of a new proton synchrotron, called the Tevatron, that is designed to reach one trillion electron volts (TeV). The Tevatron magnets have superconducting windings. The scale of the accelerators can be judged from the slight curvature of the tunnel, which has a circumference of 6.3 kilometers. (Photo by Fermi National Accelerator Laboratory.)

is a scarce resource, and it is imperative that it be used as efficiently as possible. Technical innovations have brought a substantial reduction in the cost per unit energy of accelerating a particle. It is encouraging to note that another means for minimizing the total world expenditure is now emerging: through international cooperation the unnecessary duplication of facilities can be avoided, and projects too large for any one nation can be under-

taken by regional groups of nations and perhaps eventually through a worldwide collaboration.

The large investment now being made in instruments for high-energy-physics research can be justified only because the preceding generations of accelerators have already proved their worth. Fifty years ago only two kinds of apparently indivisible particle were recognized: the electron and the proton. The remaining constituent of the atom, the neutron, was discovered in 1932. In subsequent years, through experiments with cosmic rays and with early accelerators, several additional particles were identified. One of the first was the positron, the antiparticle of the electron. Others were the neutrino, a particle without mass or electric charge, and the muon and pion, which have masses intermediate between those of the electron and the proton.

In the 1950's, when more powerful accelerators began operating, there was an unexpected and in some respects alarming proliferation in the number of known particles. Within a few years the list extended to more than 100, most of them classified as hadrons, or nuclear particles. Among the hadrons were several with the new property of matter called strangeness. Five years ago it was necessary to add another class of hadrons, bearing another whimsically named property, charm. The pace of discovery continues to increase. Particles that apparently signal the existence of two more classes have been observed. The newest classes, which have only begun to be catalogued, are distinguished by properties called truth and beauty or top and bottom.

For a time it seemed that all of these particles might have to be accorded equal status as elementary objects. That possibility was deeply troubling, as it was difficult to reconcile with the conviction that the laws of nature should be reasonably simple. It was subsequently discovered, however, that all the hadrons could be arranged in logical patterns, some of which have a lovely snowflake form. Moreover, the existence of such patterns could be understood if it was assumed that the hadrons are not elementary but are made up of the more fundamental entities that have been given the name quarks.

In the view that now prevails among physicists there are just two kinds of elementary particles: leptons and quarks. Among the leptons the most familiar particle is the electron. Also included in that class are the muon and two kinds of neutrino, one associated with the electron and one with the muon. A few years ago a new lepton was discovered and given the designation tau. Presumably the tau also has an associated neutrino, so that there should be six leptons altogether.

There also appear to be six kinds of quarks, labeled up, down, strange, charmed, top and bottom. (As yet there is no experimental evidence for the top quark, but because all the other quarks and leptons come in pairs it is assumed that the bottom quark

also has a partner.) No one has observed a quark in isolation, but there are substantial reasons for believing in their existence. Every known hadron (and there are now a few hundred) can be explained as a combination of quarks or of quarks and antiquarks, formed by explicit rules.

Cataloguing the elementary particles constitutes only half the problem of understanding the structure of matter. It is also necessary to understand the forces that hold the particles together and give rise to their motions. Four basic forces are generally recognized. In order of increasing strength they are gravitation, the weak force (which is responsible for radioactive beta decay), electromagnetism and the strong force. Gravitation and the weak force are universal: they act between all kinds of particles. Electromagnetism has a direct influence only on particles that carry an electric charge. The strong force acts only on hadrons or on their constituents, the quarks.

Each force is thought to be transmitted from point to point by the exchange of an intermediary particle. For electromagnetism the intermediary particle is the photon: the quantum of light and of other forms of electromagnetic radiation. Thus the repulsion between two electrons is described as coming about from the exchange of photons, which are emitted by one electron and absorbed by the other. (A similar repulsion would be observed between two children on ice skates throwing a ball back and forth.) The mechanism of interaction is similar for the other forces, except for the identity of the exchanged particle. The quantum of gravitation is the graviton, and the strong force between quarks is transmitted by the particles called gluons. For the weak force the mediating particle is the intermediate vector boson, now also known as the weakon; it comes in three charge states, designated W^+, W^- and Z^0. Whereas the photon, the graviton and presumably the gluons are all massless, the weakon is expected to be very heavy.

One of the most far-reaching theoretical developments of the past several years was a demonstration that the weak force and the electromagnetic force can be unified. Although the two forces are quite dissimilar in their observable characteristics, they can now be understood as manifestations of a single underlying phenomenon. The unification is precisely analogous to the 19th-century realization that electric and magnetic forces are merely different manifestations of charge. Work is now under way toward the construction of a "grand unification" that would include the strong force with the unified weak and electromagnetic forces. The ultimate source of force, however, remains as mysterious today as the source of the Nile was in the 19th century, and the search for that source is just as romantic.

However much has been learned in the past 50 years, it would be misleading to suggest that the present understanding of elementary particles is even approaching finality or completion. The status of the field is tantalizing rather than satisfying; there is no shortage of questions to be answered. A first order of business for the new accelerators will be filling in the blanks in the catalogue of hadrons, particularly those that incorporate top and bottom quarks in their structure. It is also important to find out whether the list of quarks and leptons ends with the six of each that are now known or whether more will be found at higher energies. In a sense six quarks and six leptons are already too many; all of the ordinary matter in the universe could be constructed out of just four elementary particles: the electron, the electron neutrino and the up and down quarks. The existence of the other leptons and quarks, which appear only in high-energy-physics experiments, is a puzzle.

Another puzzle is the failure of all attempts so far to detect a free-quark. Various theoretical constructs have been offered, after the fact, to explain why quarks should be permanently confined to hadrons. The possibility remains, however, that a quark can be knocked loose from a hadron if enough energy is supplied. Future experimental programs are therefore certain to include quark searches.

One of the trophies that will be hunted with the new accelerators is the weakon, the transmitter of the weak force. The three kinds of weakon are estimated to have masses roughly 100 times the mass of the proton; creating particles that heavy is beyond the capabilities of any existing accelerator, and it may not be possible for several years.

The historical development of elementary-particle physics might well be read as an extended lesson in skepticism. Over the course of the past century the realm of inquiry has progressed from the atom to the atomic nucleus to the hadrons that make up the nucleus to the quarks that make up the hadrons. Each of these objects has been considered for at least a time to be an elementary particle, one with no internal structure. It may be too soon to declare an end to the progression by assuming that the quarks (and the leptons) are truly elementary. They too could be composite objects, made up of simpler components. Indeed, one can imagine the endless regression that was visualized by Jonathan Swift:

So, naturalists, observe, a flea
Hath smaller fleas that on him prey;
And these have smaller still to
   bite 'em;
And so proceed *ad infinitum.*

Theorists are already contemplating the next step in the regression. For example, Haim Harari (see Chapter 12) of the Weizmann Institute of Science in Israel has suggested that both the quarks and the leptons could be constructed of just two kinds of "ultimate" particles. He calls them rishons, from the Hebrew word for elementary.

The rationale for employing particle accelerators to explore the structure of matter is straightforward. Smashing two objects together is merely a means for breaking them down into their component parts. If the collision could be made sufficiently violent, the particles might be reduced to their ultimate and unbreakable constituents. An interaction between high-energy particles is not entirely like an automobile accident, however. Particles not only are knocked out of the target and the projectile but also can be created anew out of the energy brought to the collision by the accelerated particle. For example, the upsilon particle, which is thought to include a bottom quark, can be made in collisions between energetic protons even though the mass of the upsilon is 10 times that of the proton.

Another way of considering the functioning of an accelerator is by analogy with a microscope. The ultimate limit to the resolution of a microscope is the wavelength of the radiation with which the specimen is illuminated. Patterns much smaller than the wavelength cannot be resolved; as a result the light microscope cannot distinguish objects smaller than about 10^{-5} centimeter. In quantum mechanics a particle of matter can be described as a wave, which has a wavelength inversely proportional to the momentum of the particle. If an accelerator is conceived as a large microscope, the motive for increasing the energy is to reduce the particle wavelength and thereby improve the resolution. The largest accelerators now operating have an effective resolution of about 10^{-16} centimeter, which is a thousandth of the diameter of the proton.

The newest accelerators exploit the same fundamental principles as the first ones. The force employed to accelerate the particles is electromagnetism, and so only particles that have an electric charge can be accelerated; they are usually either protons (with a charge of +1) or electrons (with a charge of −1). The particles are injected into a vacuum chamber, the vacuum being necessary to prevent the moving particles from colliding with air molecules. An electric field then sets the particles in motion. In the simplest case a high voltage is applied across a pair of electrodes. Electrons are propelled toward the positive electrode; protons move in the opposite direction, toward the negative electrode. A simple accelerator of this kind is found in the picture tube of a television receiver.

The standard unit of measurement for particle energy is the electron volt, abbreviated eV. One electron volt is the energy acquired by an electron when it is accelerated through a potential difference of one volt. The same unit serves to measure the energy of moving protons or of any other particles. For convenience various multiples of the electron volt are employed in specifying the energy of accelerators. The kiloelectron volt (keV) is 1,000 electron volts, the megaelectron volt (MeV) is a million, the gigaelectron volt (GeV) is a billion and the teraelectron volt (TeV) is a trillion, or 10^{12} electron volts. Because mass and energy can be freely interconverted it is customary to give the mass of a particle in terms of its energy equivalent, measured in electron volts. Thus the "mass" of the proton is 938 MeV.

In principle any energy could be achieved with a simple accelerator made up of two electrodes, merely by raising the voltage to the appropriate level. In practice the maximum potential that can be sustained across a pair of electrodes is several million volts, so that such accelerators are confined to energies of no more than several MeV. The limit is determined by the onset of arcing between the electrodes or by the breakdown insulators.

In order to reach higher energies it is necessary to accelerate a particle in stages, giving it a sequence of small pushes instead of one big push (see Figure 4.2). The most obvious way of arranging for such a gradual acceleration is by stringing together many small accelerator stages one after the next. That is the principle of the linear accelerator, or linac. Instead of applying a continuous high voltage to the electrodes in each stage, an alternating electric field is set up by an oscillator connected to each set of electrodes, forming the structure called a radio-frequency cavity. The oscillators for successive cavities are synchronized in such a way that the electric field always has the correct polarity to accelerate, rather than retard, the moving particle. In effect an electromagnetic wave travels continuously through the vacuum tube and the particle rides the electrical wave as a surfer rides a water wave.

There is only one large linac now operating. It is at the Stanford Linear Accelerator Center (SLAC) near Stanford University. Completed in 1961 at a cost of some $115 million, it is two miles long and made up of 82,560 radio-frequency cavities. The SLAC device was designed to accelerate electrons to an energy of 22 GeV; a program for replacing the radio-frequency oscillators with more powerful units is raising that energy to 30 GeV. Even without the upgrading of the machine SLAC would have remained for some time to come the most powerful electron accelerator in the world.

The limit to the maximum practical energy of a linear accelerator is the cost of the thousands of accelerator cavities and their associated radio-frequency power supplies. The way to avoid that cost is to employ only a few cavities but to make each particle pass through them many times. Under the influence of a magnetic field an electrically charged particle follows a curved trajectory. By arranging many magnets in a ring the particle can be made to follow a circular orbit, or any other closed curve. A bunch or cluster of particles can circle the ring several million times, passing through the radio-frequency cavities and gaining energy on each revolution. An accelerator built in this way is called a synchrotron.

All the large new accelerators that are now planned or under construction are synchrotrons. It is therefore worthwhile to consider their operation in somewhat greater detail. The magnets that make up the ring are of two kinds. The dipole magnets, which have two poles (one north and one south), generate a uniform magnetic field; they accomplish the bending of the particle trajectories. Quadrupole magnets, which give rise to a field with two north poles and two south poles, do not deflect the particles but focus them into a narrower beam, acting much like lenses. Interspersed among the magnets are the radio-frequency cavities, where the actual acceleration takes place. Specialized magnets and electrodes must also be provided for injecting the particles into the ring and for extracting them from it.

The synchrotron operates in cycles. When a

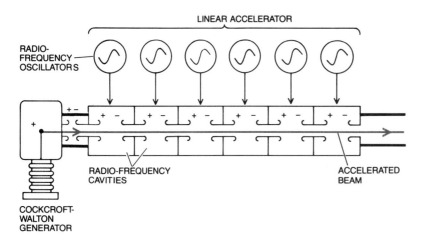

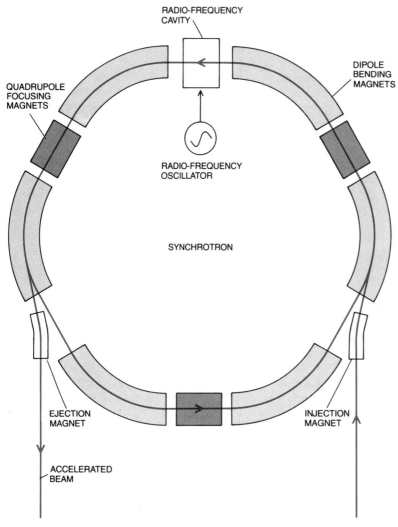

Figure 4.2 ACCELERATION OF A PARTICLE results from the force applied by an electric field. A static high voltage supplies the field in a Cockcroft-Walton generator, but the maximum energy that can be reached by this method is about a million electron volts (MeV). Higher energies call for many stages of acceleration. In a linear accelerator, many radio-frequency cavities are arranged in sequence and synchronized so that a particle receives a small push at each stage. In a synchrotron the many radio-frequency stages can be replaced by one cavity, through which the beam passes many times. Bending and focusing magnets confine the particles to a circular path.

bunch of particles is first injected, the fields of the bending magnets are adjusted so that the particles precisely follow the curvature of the vacuum tube. As the energy of the particles increases on each revolution the field strength in the bending magnets must also be smoothly increased. When the maximum energy is reached, the beam is extracted; then the magnetic field is allowed to fall to its original level in preparation for the next bunch of particles. The accelerator is called a synchrotron because the particles automatically synchronize their motion with the rising magnetic field and the rising frequency of the accelerating voltage.

The highest energies are not attempted with a single machine; instead several machines are lined up in series. Each one augments the particle energy by a factor of 10 or even 100, then passes the beam on to the next accelerator in the sequence. In several instances older accelerators serve as injectors or preliminary stages for newer and for more powerful machines.

In a proton accelerator the first stage is most often a device of the kind built in 1928 by John D. Cockcroft and Ernest T. S. Walton at the Cavendish Laboratory of the University of Cambridge. It is a large transformer and rectifier that generates a potential of about a million volts between an inner electrode and an outer shell. Protons, obtained by ionizing hydrogen atoms, are released at the inner electrode; when they emerge (through a hole in the shell), they have an energy of about 1 MeV.

The next stage is often a linac, which typically raises the energy per proton to 50 or even 200 MeV. From the linac the protons can be injected into a synchrotron, which may be the final link in the chain or may serve merely to boost the energy of the protons for injection into a larger synchrotron (see Figure 4.3).

The first large synchrotrons were the Cosmotron, built in 1952 at the Brookhaven National Laboratory, and the Bevatron, completed in 1954 at

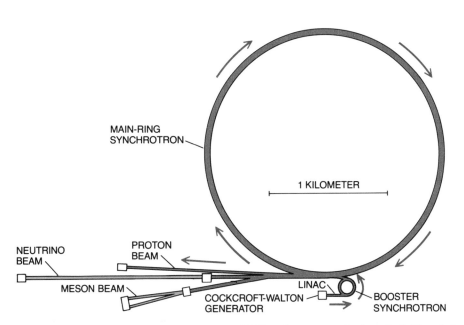

Figure 4.3 FIXED-TARGET ACCELERATOR delivers particles to an external target, where detectors and other experimental apparatus can be set up. The accelerator shown is the one at Fermilab. Protons are accelerated in four stages: a Cockcroft-Walton generator (.75 MeV), a linac (200 MeV), a booster synchrotron (8 GeV) and the main-ring synchrotron (400 to 500 GeV). Experiments can be carried out not only with the accelerated protons but also with beams of secondary particles, such as mesons and neutrinos, knocked out of the target by the impact of protons.

the University of California at Berkeley. They reached energies of 3 GeV and 6.2 GeV respectively. In design they differed from more recent practice chiefly in the disposition of the magnetic field, which provided only weak focusing of the beam. In building the next generation of larger and more powerful synchrotrons, a new system of strong focusing was introduced. The shape of the magnetic field can be described mathematically as being partly uniform (the dipole component) and partly a gradient in a direction transverse to the orbit of the beam (the quadrupole component). The quadrupole component was made stronger and was alternated in sign, so that oscillations of the particles around the desired orbit were more frequent but of smaller amplitude. As a result of this alternating gradient the aperture of the magnets and the bore of the vacuum chamber could be made smaller. It is the invention of the synchrotron and of strong focusing that has made the very large accelerators of today economically feasible.

The strong-focusing principle was first applied to synchrotrons built at Brookhaven and in Europe. The Brookhaven machine, which came to be known as the Alternating Gradient Synchrotron, or AGS, was completed in 1961 and eventually reached an energy of 33 GeV. The AGS has had a distinguished career. It was an experiment with the AGS that first revealed the existence of two kinds of neutrino, one for the electron and one for the muon. In 1974 an AGS experiment was one of two that discovered the particle labeled J or psi, which provided the first evidence of charm. The AGS is still operating and, as I shall explain below, there are plans for its further utilization.

The European strong-focusing synchrotron has turned out to be part of a much larger endeavor, what is indeed now the largest of the world's high-energy-physics laboratories. The project was undertaken by a consortium of European nations, which formed the Conseil Européen pour la Recherche Nucléaire, or CERN. (The name has since been changed to European Organization for Nuclear Research, but the abbreviation remains.) It is a model of international cooperation, which has broken new ground in solving the problems of language, money, custom and national self-interest. Today there are 12 member nations. The laboratory itself now straddles the French-Swiss border a few miles north of Geneva.

The Proton Synchrotron, or PS, at CERN was completed about a year ahead of the Brookhaven AGS. The PS too has made distinguished contributions. Most notably it was the instrument with which CERN physicists discovered a new aspect of the weak force, called the neutral weak current, which provided the first evidence supporting the unification of the weak and electromagnetic forces.

Research is still carried out with the proton beams of the PS, but what is now probably the most important function of that machine is to serve as the injector for a still larger accelerator, the Super Proton Synchrotron, or SPS. The initial plans for the SPS called for a 300-GeV accelerator to be built somewhere in Europe other than the CERN site. In 1965, however, France made available a tract of land adjoining the original laboratory in Switzerland, so that the SPS could be built next to the PS. As it happened, there was no need to disturb the surface over most of the area available. The tunnel for the SPS, which is almost seven kilometers in circumference, was bored underground with a mining machine, at an average depth of 40 meters. The beam comes to the surface only in the experimental halls.

When the SPS began operation in 1976, its energy was not 300 GeV but 400, and it has the capability of reaching 500 GeV. There is no question that the availability of the PS as an injector speeded construction and made the higher energy economically feasible. The highest-energy protons at CERN pass through five accelerators: a Cockcroft-Walton generator (550 keV), a linac (50 MeV), a booster synchrotron (800 MeV), the PS and the SPS.

At about the same time planning began for the SPS, an American project of comparable scale was undertaken. After considerable bickering among states competing for the new laboratory, a site was chosen in 1967, on farmland 30 miles west of Chicago. The facilities built there are now called the Fermi National Accelerator Laboratory, or Fermilab. It does not have the cachet of international sponsorship, but at least it is interscholastic: the operating authority is an association of 53 universities. Of course, physicists of many nations do experiments at the laboratory.

Work at Fermilab proceeded faster than at CERN, and the accelerator began operating in 1972. The original plan had called for a 200-GeV proton synchrotron, but it was possible to build a 400-GeV machine for the same cost. By 1975 the maximum energy had been raised to 500 GeV, although the cost of electric power forbids extended periods of operation at that energy.

The system of accelerators at Fermilab follows what by now should be a familiar pattern. Protons are given an initial push in a Cockcroft-Walton generator (750 keV), then pass through a linac (200 MeV) and a booster synchrotron (8 GeV) before being inserted into the main-ring synchrotron. The main ring is a little smaller than that of the SPS; its circumference is 6.3 kilometers. It is made up of 774 bending magnets, each 20 feet long, and 180 smaller focusing magnets. A series of radio-frequency cavities add 2.8 MeV to the protons' energy on each revolution. Going from 8 GeV to 400 GeV therefore requires 140,000 turns, which are completed in about three seconds. Controlling the rapidly changing events in such a large and complicated machine requires a battery of three large digital computers and many smaller satellite computers.

The process whereby a prima ballerina accelerator becomes a mere member of the ballet company, serving as the injector for a larger device, is about to be repeated at Fermilab. A new ring of magnets and radio-frequency cavities is being installed under the existing main ring, so that there will be two accelerators in the one tunnel (see Figure 4.1). Protons will be brought up to an energy of about 150 GeV in the main ring, then transferred to the lower ring for further acceleration. The new ring is made up of superconducting magnets, which can reach a field strength twice that of the old magnets. The maximum energy of the accelerator is thereby doubled, to 1 TeV. In recognition of this milestone the new accelerator has been named the Tevatron.

The use of superconducting magnets was contemplated when plans were first made for Fermilab, but the risk of an unproved technology seemed too great then. Building the Tevatron has turned out to be a challenging task even today. The most obvious problem is that of cooling almost 1,000 magnets, strung out over 6.3 kilometers of tunnel, down to 4.5 degrees Kelvin, the temperature where the special conductors of the magnet windings lose all resistance to the flow of electricity. In order to maintain that temperature a river of liquid helium will be pumped through the ring. Twenty-four small refrigeration plants will be spaced around the tunnel, and the central helium liquefier will be the largest in the world, with a capacity of 4,000 liters per hour.

The windings in the magnets are formed of a niobium-titanium alloy embedded in a copper matrix. Almost 19,000 miles of this cable will be required to complete the ring, qualifying Fermilab for another superlative: it is the largest consumer of superconducting materials in the world. At maximum field strength the superconductors will carry a current of 4,600 amperes, and when a magnet is "quenched," or loses the superconducting property, the energy stored in the field (about half a million joules per magnet) must be dissipated without destroying the windings.

A particularly taxing problem has been the need to maintain the uniformity of the magnetic field to an accuracy of better than one part in 1,000. Because the windings are immersed in their own field they are subjected to a reactive force of about one ton per linear inch. The coils cannot be allowed to move even one thousandth of an inch, however, because the movement would distort the field and might also give rise to too much frictional heat. The windings are immobilized by laminated collars of stainless steel. The alignment of the magnets is also complicated by thermal contraction when the ring is cooled to its operating temperature; a magnet six meters long contracts by about two centimeters.

There is now substantial confidence that these problems have been understood and solved. A string of 24 superconducting magnets has been installed in one segment of the tunnel, and a beam at 90 GeV has been diverted through it from the main ring. The magnets performed as expected. Completion of the full lower ring is now expected toward the end of 1981, and protons at 1 TeV should be delivered to the experimental areas in 1982. The rate of construction depends mainly on the rate of funding.

It is worthwhile pausing to consider just how much energy 1 TeV per proton is. In units more commonly applied to macroscopic objects, a 1-TeV particle has an energy of 1.6 ergs, which is roughly the kinetic energy of a flying mosquito. At full intensity the Tevatron will accelerate 5×10^{13} protons at a time, which will give the total beam an energy of eight million joules. That is comparable to the energy of a 100-pound artillery shell. If the beam should ever go out of control, it could melt the walls of the vacuum chamber and destroy the surrounding magnets; obviously such an accident must be avoided.

When the particles in a synchrotron have reached their full energy, they are nudged out of their orbit by a special magnet and deflected into an external beam line. Eventually they strike a target. Interactions of the protons with the target can be studied directly, and it is also possible to create beams of secondary particles knocked out of the target. At Fermilab, for example, there are separate areas for experiments with protons (the primary particles), with mesons (particles of intermediate mass, such as the pion), with neutrinos and muons, and with photons. Neutrino experiments have been exceptionally rewarding in recent years because the neutrino is subject only to the weak force; the properties of that force can therefore be observed without interference from other kinds of events.

The detectors and other apparatus required for the experiments are often built on a grand scale, comparable to that of the accelerators. The armor plating of two decommissioned battleships has been incorporated into the shielding of a detector at Fermilab. A detector now installed at CERN employs many tons of fine Carrara marble. These detectors are "counters": they sense the energy and direction and penetrating power of various particles and record them electronically for later analysis. Another kind of detector is the bubble chamber, where the passage of an electrically charged particle leaves a telltale track of small bubbles. From photographs of the tracks the sequence of events following a particle collision can be reconstructed. Fermilab has a bubble chamber 15 feet in diameter; CERN has three chambers, the largest about 12 feet in diameter.

The CERN and Fermilab machines are both proton accelerators, but electrons can also serve as the working medium of an accelerator. The physical principles are exactly the same, although the characteristics of the machine are somewhat different. The main reason for the difference is the greater importance in electron accelerators of the electromagnetic radiation that dissipates the energy of an accelerated particle.

It has been understood for more than a century that any accelerated electric charge must radiate electromagnetic waves. "Acceleration," in this context, refers not only to a change in the speed of a particle but also to any change in its direction. If it is made to follow a circular path, then it undergoes a continuous acceleration even if the speed is constant. The radiation emitted under these conditions is called synchrotron radiation, because it was in the synchrotron that it was first observed.

The synchrotron radiation appears as an intense beam of high-energy electromagnetic waves, with a continuous spread of wavelengths extending to the ultraviolet and X-ray regions of the spectrum. The energy drained away by this process must be made up by supplying additional radio-frequency power. Thus synchrotron radiation acts as a resistive force analogous to friction.

The energy emitted in the form of synchrotron radiation varies inversely as the mass of the particle raised to the fourth power. Because the mass of the proton is 1,836 times that of the electron the problem of supplying the lost energy is greater for electrons by a factor of 10^{13}. At the beam energies that have been attained so far synchrotron radiation is not a significant consideration in the design of proton accelerators, whereas it is the principal restraint on the energy of electron accelerators.

One solution to the problem of synchrotron radiation is to build a linear accelerator, where there is no curvature and where the radiation resulting from the gentler straight-line acceleration of the electrons is negligible. It was for this reason that a linac was chosen for the SLAC electron accelerator. The principal drawback of this approach is again the cost of the many radio-frequency cavities and the oscillators needed to excite them. In a linac an electron passes through each cavity only once, which is a great extravagance. SLAC cost $115 million; Fermilab cost almost $250 million, but the cost per electron volt was more than 10 times less than it was at SLAC.

The alternative to a linac is building a synchrotron for electrons and accepting the cost of synchrotron radiation. A compromise can be reached between construction costs and the continuing expense of operating the radio-frequency power supplies. The energy radiated per revolution is inversely proportional to the radius of curvature of the particle track, and so the energy loss declines as the accelerator is made larger.

In 1965 a 10-GeV electron synchrotron was built at Cornell University in a tunnel dug under an athletic field. At the time the machine was notable for its size; the circumference is about 630 meters. It reached about half the energy of SLAC at about a tenth the cost.

Another electron synchrotron, with a beam energy of 6.3 GeV, was constructed in 1962 in Cambridge, Mass. It has since been dismantled. A third was built at Hamburg in West Germany; it is called DESY, standing for Deutsches Elektronen-Synchrotron. With a beam energy of 7 GeV, DESY has now become the basis for a diversified high-energy-physics laboratory.

For a given radius of curvature the energy lost to synchrotron radiation increases as the fourth power of the beam energy. The energy loss becomes the dominant consideration in the design of the accelerator at about 10 GeV, and it imposes an almost impenetrable barrier beyond a few hundred GeV. Of course, the feasibility of any instrument is a matter of judgment and is subject to change; 25 years ago the limit of electron synchrotrons was thought to be about 1 GeV.

It should be said in defense of synchrotron radiation that it is not entirely wasted energy. One effect of the radiation is to damp out small excursions of the electrons away from their mean trajectory, making a beam of electrons easier to control than a beam of protons. What is more important, the radiation itself has become a valuable tool for biological and materials studies. It is the most intense source known of ultraviolet radiation and X rays. Facilities for exploiting the radiation have been set up at several high-energy-physics laboratories, and a number of small accelerators have been built explicitly as sources of synchrotron radiation.

In spite of the vexing problem of synchrotron radiation, electron accelerators will have an increasingly important part in future high-energy-physics programs. The reason is that the electron is a much simpler particle than the proton, and therefore makes a better probe of the structure of matter. When two protons (or other hadrons) collide, the outcome is complicated by the multiquark structure of the particles. The electron gives no evidence yet of having an internal structure, and so the results of electron interactions are more easily interpreted. This purity is achieved in double measure when an electron collides with another electron or with a positron. Such collisions can be arranged by building a device in which an accelerated beam impinges not on a fixed target but on another beam of particles.

If the entire 500 GeV per particle that is generated by the largest proton synchrotrons were released in a collision with a fixed target, many of the goals set for the next generation of particle accelerators would already have been achieved. The weakon, for example, is expected to have a mass near 100 GeV, and so it would have been seen by now. Not all the energy of an accelerated particle is made available for the creation of new particles, however, when a moving projectile collides with a fixed target. Instead a large fraction of the energy goes into setting in motion the system made up of the two particles.

Consider a 500-GeV proton striking a stationary proton, such as the nucleus of a hydrogen atom. If the projectile were stopped by the collision, then all the energy it had acquired would have to be dissipated. What actually happens is that the accelerated proton keeps moving and the target proton moves in the same direction along with it. Indeed, because of relativistic effects the rapidly moving accelerated proton has a mass 530 times that of the target proton. The accelerated particle is not stopped by the collision, just as a truck is not stopped when it collides with a bicycle. The two particles together retain an energy of 469 GeV, and only 31 GeV is made available for the creation of new particles.

The energy released in a particle collision can be calculated most easily by changing one's frame of reference. The collision is seen most realistically by an observer who is moving parallel to the beam and at the same speed as the center of mass of the two-particle system. Such an observer would see the beam proton and the target come together symmetrically with an energy of 15.5 GeV each, or a total energy of 31 GeV. That is the center-of-mass energy. According to the kinematics of relativity theory, the center-of-mass energy grows only in proportion to the square root of the beam energy, so that as accelerators become larger more of the energy invested is in this sense wasted. In order to reach a center-of-mass energy of 100 GeV with a fixed-target proton accelerator the beam energy would have to be more than 10,000 GeV (see Figure 4.4).

The view of the collision from the center-of-mass frame of reference suggests a solution to this problem. If the two particles are both accelerated to the same energy but are moving in opposite directions, they can be made to collide head on. In the simplest analysis both could be stopped in their tracks. Then all their energy would be liberated. If the particles have the same rest mass and if they have been accelerated to the same energy, the center-of-mass energy is simply the sum of the two beam energies.

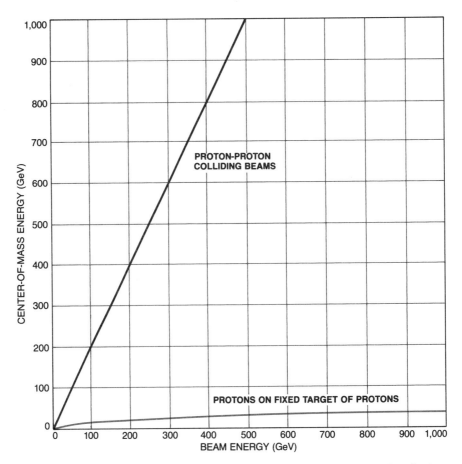

Figure 4.4 EFFECTIVE COLLISION ENERGY is much greater in colliding-beam devices than fixed-target accelerators. The energy available for creating new particles is the energy of the collision when the collision is viewed in the frame of reference of the center of mass of the colliding particles. In a storage ring the center-of-mass energy is simply twice the beam energy. In a fixed-target accelerator at low energy the center-of-mass energy is proportional to the square root of the beam energy. Fixed-target accelerators have the redeeming property of higher luminosity and yield a variety of secondary beams, which cannot be generated by a colliding-beam device.

In order to reach 100 GeV all that is needed is a pair of beams with energies of 50 GeV each.

Head-on collisions can be arranged by building storage rings, in which beams of particles circulate continuously. A storage ring resembles a synchrotron: it has an annular vacuum chamber, which is surrounded by bending and focusing magnets, and has at least one radio-frequency cavity. Usually the ring is employed not to increase the particles' energy but to maintain them in a constant orbit and at a constant energy. The radio-frequency cavity supplies only enough power to make up for the energy lost by synchrotron radiation.

One method of arranging head-on collisions is to build two storage rings that are tangent to each other. The particles collide at the point of contact. In another plan two rings are interlaced, each of them following an undulating path like the strands of a braided hoop; the beams collide at each intersection. The cleverest of the storage-ring designs is possible only when the beams being stored are made up of particles and their corresponding antiparticles, such as electrons and positrons. It is then possible to make do with a single ring. Suppose the magnetic fields and the radio-frequency cavities are adjusted to maintain electrons circulating clockwise at some fixed energy. Positrons, having the opposite electric charge, respond to magnetic and electric fields in a manner exactly opposite to that of electrons. The positrons can therefore be injected into the same ring, but in a counter-clockwise direction, and the single set of magnets and radio-frequency cavities will maintain both kinds of particle in orbit. If the ring holds one bunch of electrons and one bunch of positrons, collisions will be observed at two points diametrically opposed (see Figure 4.5).

Given the enormous energy advantage of a colliding-beam device, it may seem surprising that anyone would now contemplate building a fixed-target accelerator. Energy, however, is not the only pertinent measure of an accelerator's performance. Another important factor is the rate at which interactions are observed. A solid or liquid target has a far greater density of particles than an accelerated beam, with the result that the interaction rate is much higher in a fixed-target accelerator. In the colliding-beam device, most of the time the particles in the beams do not collide but pass through one another without interacting at all. The probability of interactions is proportional to a parameter called the luminosity, which essentially measures the brightness of the beam. The luminosity of fixed-target machines is greater than that of storage rings by a factor of about a million.

Interpreting the results of a physics experiment requires a statistically significant sample of events. Collecting such a sample can be difficult when the events of interest take place at a rate of only a few per day. For this reason storage rings may be expected to give the first glimpse of the commonest phenomena at high energy, but greater accuracy and the observation of rare events may follow only when the fixed-target accelerators reach comparable center-of-mass energies.

Another advantage of the fixed-target accelerator is versatility. A storage ring generally provides only one kind of collision, such as protons on protons or electrons on positrons. A fixed-target machine, on the other hand, can generate a variety of secondary beams, including neutrinos, muons, pions and other mesons, antiprotons and the massive particles called hyperons. Indeed, the small fraction of the beam energy that is made available in the center-of-mass system is an advantage for creating secondary

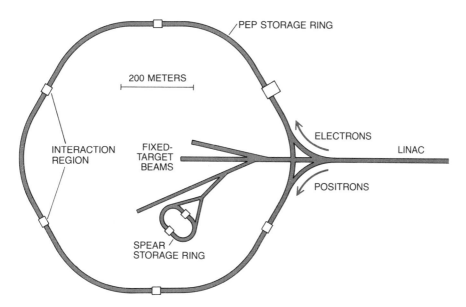

Figure 4.5 COLLIDING-BEAM DEVICES store particles at high energy and bring them into collision head on. The storage ring shown is called PEP and is at the Stanford Linear Accelerator Center (SLAC). Electrons and positrons from the two-mile linear accelerator at SLAC will be injected into the PEP ring in opposite directions and main-tained at an energy of up to 18 GeV. Collisions between the counter-rotating beams will take place in six interaction zones around the circumference, where detectors can be built around the beam pipe. Smaller SPEAR ring has been operating since 1972 at center-of-mass energies of up to 8 GeV.

beams; it is the large quantity of "wasted" energy that accelerates the secondary particles.

One further advantage of the fixed-target accelerator has already been described at length: the machine built to supply particles for research can later supply them to a larger accelerator, or to a storage ring.

The first electron-positron storage ring was begun in 1959 by Bruno Touschek and an enthusiastic group of Italian physicists at the National Laboratories of the National Committee for Nuclear Energy (C.N.E.N.) at Frascati near Rome. Called ADA, for Annello d'Accumulazione, the ring was later moved to the Orsay laboratory outside Paris. The energy of each beam was .25 GeV, yielding a center-of-mass energy of .5 GeV.

By the early 1970's half a dozen other storage rings were in operation. ADA was the direct progenitor of a larger ring at Orsay (a third has since been built there) and of a ring at Frascati that was named ADONE, for "Big ADA," with an energy of 1.5 GeV per beam. The first in a succession of pioneering storage-ring projects at Novosibirsk in the U.S.S.R. was completed in 1965. The Cambridge Electron Accelerator was modified at about that time to form a storage ring with an energy of 3.5 GeV per beam. The life span of the Cambridge machine was brief, but the machine revealed an anomaly in the total rate of electron-positron interactions that presaged the discovery of charm.

By 1974 storage rings at two laboratories had exceeded 4 GeV per beam. One of them, called SPEAR, is at SLAC and is filled with electrons and positrons from the linac there (see Figure 4.5). The other, called DORIS, was built next to the DESY synchrotron in Hamburg, and receives its particles from that source.

It was with the SPEAR storage ring that the first charmed particle, the J or psi, was seen at 3.1 GeV. As I have mentioned, the same particle was discovered simultaneously in a fixed-target experiment at Brookhaven, but the signal at SPEAR was much clearer. Workers at DORIS were able to detect the new particle immediately when they tuned their instrument to the right energy. ADONE had just missed seeing the psi because it was designed for a maximum center-of-mass energy of exactly 3 GeV, but it was possible to nudge the machine over this limit and detect the particle. The subsequent exploration of the spectrum of charmed particles was carried out largely with these three storage rings.

The discovery of charm revitalized elementary-particle physics. Obviously it pleased the experimenters who had made the discovery, but it was also rewarding to theorists, who had predicted the existence of the charmed quark and had waited 10 years for it to be revealed. After such a success it is little wonder that plans were immediately drawn up for a new round of electron-positron storage rings. A proposal to build a larger ring at SLAC, in collaboration with the Lawrence Berkeley Laboratory, was soon approved; it inherited the name PEP. In Europe two quite similar projects were considered, one in Hamburg to be called PETRA and the other at the Rutheford Laboratory in England. The two proposals were brought before an international body, the European Committee for Future Accelerators; following the recommendation of the committee, construction began on the German project but the Rutherford plan was dropped.

The specifications of PETRA and PEP are remarkably similar. PETRA is designed to reach a maximum energy of 19 GeV per beam, PEP 18 GeV. In size (2.2 kilometers in circumference) and luminosity they are almost identical. Most likely the costs are also similar, although differences in accounting methods make the comparison difficult.

As soon as construction began it was apparent the two laboratories were in a race. PETRA has won: it began operating in 1978, whereas PEP is not expected to be in operation until early this year. The faster progress of PETRA can be attributed to the elegant but spare design of Gustav Adolph Voss, to the fact that the German group had a head start and perhaps also to the fact that money was made available as it was needed. Indeed, the builders of PETRA were encouraged to spend quickly in order to stimulate business in the Hamburg area. PEP has been nursed along on funds doled out almost a dollar at a time, and it has also been slowed by heavy rains (which flooded the tunnel excavations) and by labor trouble.

The first experiments at PETRA suggest that both machines have an interesting future. Preliminary data provide encouraging support for quantum chromodynamics, the theory that has been devised to describe the interactions of quarks and gluons inside hadrons.

The race to exploit the simplicity of electron-positron collisions at higher energy is not being contested by PEP and PETRA alone; a dark horse has emerged. It is the Cornell Electron Storage Ring, or CESR, which has been squeezed into the same tunnel with the existing Cornell electron synchrotron. Under the direction of Boyce D. McDaniel of Cornell the project was completed in less than two years and at a cost of less than $20 million. The first electron-positron collisions were detected last summer, and experiments have now begun. CESR is unusual in being sponsored by the National Science Foundation; virtually all other high-energy-physics laboratories in the U.S. are funded by the Department of Energy.

A n innovative method of filling the storage ring has been introduced at CESR by Maury Tigner of Cornell. Bunches of particles are passed back and forth between CESR and the electron synchrotron, which have different radii. By taking advantage of the slight delay in the outer ring, many bunches can be superposed in the storage ring, thereby improving the luminosity of the beams. The maximum energy is 8 GeV per beam, which may turn out to have been a fortunate choice. It is intermediate between the energies of SPEAR and DORIS and those of PEP and PETRA, and it falls in a range where a rich spectrum of new particles is expected to be found. Those particles are the ones related to the upsilon, incorporating the bottom quark.

Heady with success, the Cornell group has made an initial design for an electron-positron storage ring capable of 50 GeV per beam. With a center-of-mass energy of 100 GeV, a particularly exciting prospect is the copious production of the neutral weakon, or Z^0. At that energy the electrons and positrons would radiate 1.6 GeV of synchrotron radiation per turn; this energy loss would be made up by superconducting radio-frequency cavities with a total power consumption of 36 megawatts. The machine would be four kilometers in circumference, and Tigner estimates it would cost $120 million.

A still grander electron-positron device is already in an advanced stage of design. Proposed by physicists at CERN, it is called LEP, for large electron-positron storage ring. Several candidates for the next large-scale European high-energy-physics facility were submitted to the European Committee for Future Accelerators in 1978. LEP won the endorsement of the committee, and a detailed design study was therefore undertaken. No funds have yet been committed to the construction of the ring, but if money is made available for a large European project, LEP will presumably have priority.

In an initial stage the maximum energy of LEP might be about 86 GeV per beam, but that could later be raised to as much as 130 GeV by adding superconducting radio-frequency cavities. In order to minimize the energy loss through synchrotron radiation the ring would be made very large, with a circumference of 30 kilometers. As with the SPS, the tunnel would be excavated from underground. Indeed, it would extend under the Jura Mountains west of Geneva, and three of the experimental halls would be half a mile underground. By locating LEP adjacent to the SPS, facilities could be added later for collisions between electrons in one ring and protons in the other. Indeed, once the LEP tunnel is in place a ring of superconducting magnets for a proton synchrotron could be added to it. Protons injected by the SPS could be brought to an energy of between 3 and 6 TeV, although this potentiality has not yet been publicly mentioned by the proponents of the machine.

The estimated cost of the initial phase of LEP is 1,275 million Swiss francs (some $800 million), more than the combined cost of Fermilab and the CERN SPS. If construction were started soon, the ring might be operating by the late 1980's.

T here is no reason for colliding-beam devices to be confined to electrons and positrons. In fact, a proton-proton collider has been operating at CERN since 1971. Called the Intersecting Storage Rings (ISR), it consists of two interlaced rings that store counter-rotating proton beams. The rings cross over at eight points around their circumference, meeting at a shallow angle. Detectors can be set up in seven of the interaction zones.

The protons stored in the ISR are supplied by the adjacent PS, at energies of up to 28 GeV. In the ISR they can be accelerated slightly, to an energy of 31 GeV per beam. The resulting center-of-mass energy, 62 GeV, is the highest available today. To reach the same center-of-mass energy with a fixed-target accelerator would require a beam energy of 2 TeV.

Even higher center-of-mass energies may be achieved at Fermilab and at CERN, although the luminosities will be much lower. The extreme energies would be obtained through improvised schemes for operating the large synchrotrons as storage rings. At Fermilab the presence of two rings in the same tunnel makes possible a variety of colliding-beam arrangements. One plan would be to store counter-rotating beams of protons in the two rings and to bring them into collision at one of the straight sections that are spaced around the rings. The Tevatron could be operated at its full 1-TeV energy as a storage ring, but the maximum steady magnet current would limit the main ring to about 250 GeV. The center-of-mass energy for such collisions is about 1 TeV. By pulsing the main ring up to its full energy a 40 percent improvement in center-of-mass energy could be obtained, although at considerable cost to the rate of interactions.

A proposal that would yield even greater energy and one that, for the present at least, is being seriously pursued, is to employ the Tevatron alone as a single-ring colliding-beam device for protons and antiprotons, operating on the same principle as the electron-positron rings (see Figure 4.6). The difficulty of such a plan is in accumulating a sufficiently

intense beam of antiprotons, which are not available in ordinary matter. Antiprotons are formed in collisions of protons with a fixed target, but they emerge with a wide range of speeds and directions and cannot in that state be injected into the narrow aperture of a synchrotron. Viewed in the antiprotons' own frame of reference the rough beam of them resulting from the collisions is a hot gas, the temperature of which must be reduced substantially before the particles can be employed in a colliding-beam experiment.

The prospects for building a proton-antiproton storage ring depend on two recently developed methods for artificially cooling a beam of antipro-

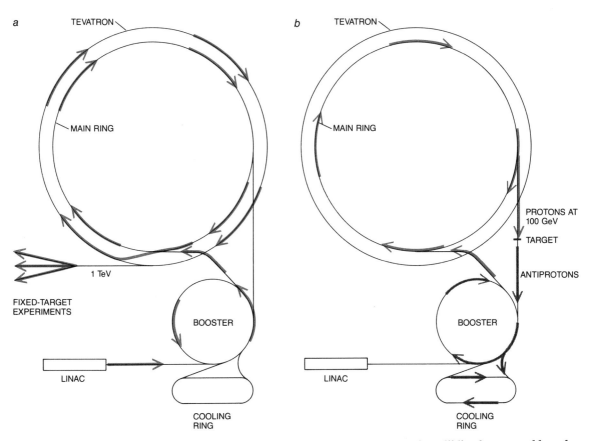

Figure 4.6 VERSATILITY OF THE MULTIPLE RINGS at Fermilab will enable the facility to operate either as a fixed-target accelerator or as a colliding-beam device. Protons will be delivered to external targets (*a*) either by the main ring (at energies of up to 500 GeV) or by the Tevatron (at 1 TeV). One scheme for colliding beams would employ the Tevatron to store counter-rotating protons and antiprotons. Protons would first be brought up to about 100 GeV in both synchrotrons; they would then be extracted from the main ring and employed to generate antiprotons (*b*),

tons (see Figure 4.7). One method, called electron cooling, was invented by the late Gersh I. Budker and his colleagues at Novosibirsk. A rough beam of antiprotons is first confined in a low-energy storage ring of very large aperture. Electrons are then passed through a long straight section of the ring in such a way that they move parallel to the average path of the antiprotons and at the same average speed. The electrons have a much lower temperature, and as a result of collisions they carry off the randomly directed components of the antiprotons' momentum. In effect the hot gas of antiprotons gives up its heat to the cold gas of electrons. At the end of the straight section the two kinds of particles are separated by a magnetic field. The electrons

strike a collector, but the antiprotons continue to circulate and pass through the cooling region repeatedly.

Electron cooling was demonstrated by Budker, and larger cooling rings have been set up both at CERN and at Fermilab. The CERN workers were the first to repeat the Novosibirsk results; they found the rate of cooling was even faster than had been expected. The Fermilab plan calls for injecting both protons and cooled antiprotons into the Tevatron at an energy of about 150 GeV. A rather complicated procedure of shuttling particles among the main ring, the Tevatron, the booster and the cooling ring will be required in order to inject both beams. The particles and antiparticles will then be accelerated

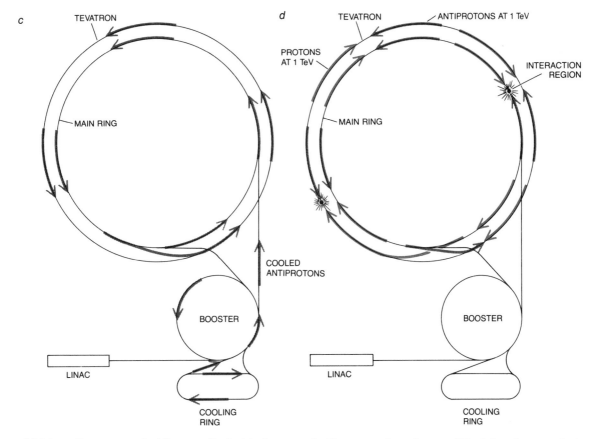

which have the same mass but the opposite electric charge. The antiprotons would be shunted through the booster to a "cooling" ring, where they would be formed into a well-collimated beam. When enough antiprotons had been accumulated, they could be returned through the booster to

the Tevatron, where they would be injected counterclockwise, opposite to its usual sense of rotation (c). Protons and antiprotons would then be accelerated simultaneously to 1 TeV, yielding a center-of-mass energy of 2 TeV(d).

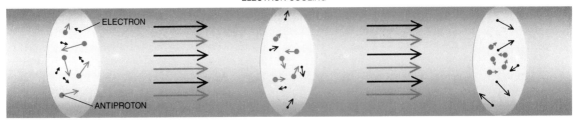

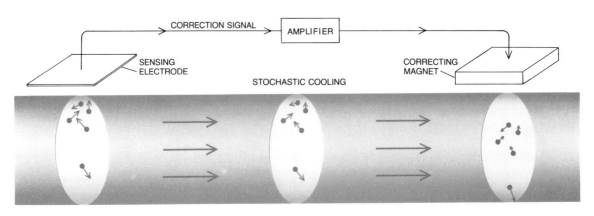

Figure 4.7 BEAM-COOLING TECHNIQUES are essential to the success of proton-antiproton storage rings. The antiprotons are created with a comparatively wide range of speeds and directions and therefore cannot be injected directly into an accelerator. When the protons are viewed in their own frame of reference, they make up a hot gas, the particles of which have randomly directed velocities. One means of cooling the gas employs electrons to carry off the heat. In repeated collisions with the antiprotons the electrons come away with most of the randomly directed momentum. In another cooling method (stochastic cooling) the average position of the beam is sensed, and then a correction is applied to keep the beam centered in the vacuum tube. Some particles may be affected adversely by the correction, but a majority will always have their extraneous components of motion reduced. After many repetitions of this process the antiproton beam will be cooled.

simultaneously to any energy up to the maximum of 1 TeV. Thus the center-of-mass energy can reach 2 TeV.

At CERN another method of antiproton cooling is being tested. Invented by Simon van der Meer of CERN, it is called stochastic cooling. The antiprotons are again stored in a small ring of large aperture, this one equipped with a system of electronic sensors and a special orbit-correcting magnet. On each revolution the sensors detect the average position of the particles at some point along the ring. If the center of charge of a fluctuation of the particles in the beam has strayed from the axis of the vacuum chamber, a correction signal is calculated and dispatched to the correction magnet. (Because the antiprotons move at nearly the speed of light the signal can reach the magnet before the particles do only by taking a short cut across a chord of the circle.) The correction magnet restores the center of charge to its proper position. In the process some particles may well be further deflected from their proper trajectories rather than returned to them, but the majority always move in the proper direction. The system of sensor and magnets acts like Maxwell's demon, cooling the beam by deft manipulations of the distribution of particle velocities.

At CERN the protons and antiprotons will be stored in the SPS (see Figure 4.8). Because of limits on the steady-state current in the conventional magnets, the maximum sustained energy will be 270 GeV per beam. That still amounts to 540 GeV in the center-of-mass system. By pulsing the magnets a center-of-mass energy of 1,000 GeV could be reached, but at a much lower average rate of interactions.

The parallel efforts of CERN and Fermilab to establish proton-antiproton collisions will undoubtedly be seen as another race, and with some justice. One prize may well be the first observation of the weakon. CERN has committed more of its resources to the project and for now has a considerable lead, but the estimated completion dates, early in 1981 for CERN and one or two years later for Fermilab, are close enough for factors unforeseen to determine the outcome. In any case it must also be said that a spirit of

generous cooperation prevails between the two laboratories. Both of the projects were largely inspired by the brilliant conceptions of the same man: Carlo Rubbia of Harvard University.

Neither of these makeshift storage rings is expected to have a high luminosity. The facilities for experimental apparatus will also be less than ideal. Furthermore, there will be fierce competition for running time with the very attractive fixed-target programs, which are incompatible with the colliding-beam experiments. The one irresistible allure of the beam-cooling projects is the opportunity to reach extremely high energy quickly and cheaply.

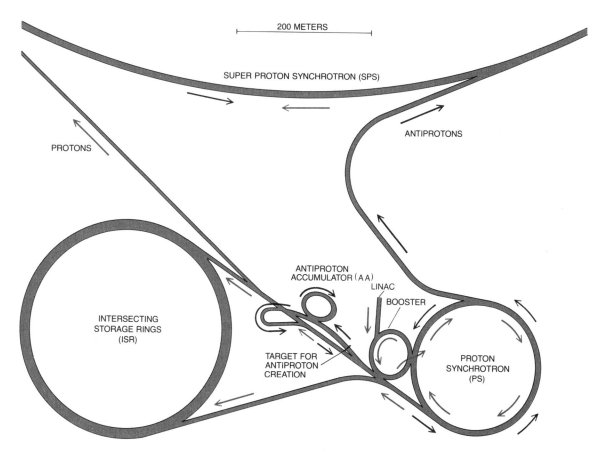

Figure 4.8 CONVERSION OF A SYNCHROTRON is planned at CERN. A proton-proton colliding-beam facility, the Intersecting Storage Rings (ISR), consists of two interlaced rings, with eight crossover zones where collisions can be observed at center-of-mass energies of up to 62 GeV. A project now under way will employ the much larger ring of the Super Proton Synchrotron (SPS) for proton-antiproton collisions. The source of protons for all the CERN devices is the Proton Synchrotron (PS). Antiprotons will be created as secondary particles from the PS beam and collected in a cooling ring called the Antiproton Accumulator (AA). Protons and antiprotons will then be injected in opposite directions into SPS and brought up to 270 GeV each. Here maximum center-of-mass energy will be 540 GeV. Only a small arc of SPS is shown.

What the CERN and Fermilab colliders will provide is a glimpse from Mount Pisgah of the promised land. Will it be LEP that will actually take us there?

The impression should not be given that the only interesting high-energy-physics laboratories are in the U.S. and Western Europe. In Japan a 12-GeV proton synchrotron, KEK, has been operating for several years and a much larger device, TRISTAN, is planned. The Chinese have announced plans for an ambitious complex of accelerators. The scope of the accelerator laboratories in the U.S.S.R. is comparable to that of the American and European programs. A series of four electron-positron colliding-beam devices has been built at Novosibirsk, the latest one having a beam energy of 7 GeV. There is a 6-GeV electron synchrotron at Yerevan in Armenia, and there are small proton synchrotrons at Moscow (7 GeV) and at the Joint Institute for Nuclear Research in Dubna (10 GeV). The Serpukhov Institute for Nuclear Physics, 60 miles south of Moscow, has an accelerator that for several years was the most powerful in the world: it is a 76-GeV proton synchrotron. When a proposed expansion of the facilities there is completed, the Serpukhov laboratory may reclaim that distinction.

In the new project, which is called UNK, protons will be transferred from the 76-GeV accelerator to a new site six kilometers away. There the protons will be injected into a 400-GeV synchrotron, which will serve as the booster for a third synchrotron, to be built in the same tunnel, with a maximum energy of 3 TeV. The booster will be of conventional design, but the 3-TeV ring will employ superconducting magnets. The magnets are now being developed at Serpukhov and at the Saclay Nuclear Research Center outside Paris.

Multiple modes of operation are envisioned. The 3-TeV protons and secondary beams derived from them will be available for fixed-target experiments. Collisions between the 3-TeV protons and the 400-GeV protons from the booster will also be possible, yielding a center-of-mass energy of somewhat more than 1 TeV. The possibility has also been discussed of adding of 20-GeV electron synchrotron, which would provide for electron-proton collisions. Beam cooling would supply antiprotons for collisions with a center-of-mass energy of 6 TeV. If a second large ring were constructed, the same energy could be attained in proton-proton collisions. Plans and specifications for UNK are now being drawn up; the machine will take some seven years to complete.

Figure 5.1 shows the location of many of the accelerator complexes across the globe.

A dream that has not yet even begun to make its way through the bureaucracy is the Site-Filler or Pentevac accelerator that could be built at Fermilab. The idea began with the question: What is the most powerful accelerator that could be fit into the Fermilab site? The land available will accommodate an accelerator no larger than five kilometers in diameter, or 17 kilometers in circumference. The energy that could be developed in a proton synchrotron of that size depends strongly on future developments in the technology of superconducting magnets. The maximum field strength possible now in a magnet suitable for an accelerator is about 50 kilogauss. It seems reasonable to assume that in about 10 years with improved materials and improved methods of fabrication the limit will be raised to 85 kilogauss (although eventually much higher fields will be attained). In that case the accelerator could reach 5 TeV. A varied menu of experiments would then be made available. Included on the list would be fixed-target beams, proton-proton collisions at energies of up to 4.5 TeV and proton-antiproton collisions at 10 TeV. An electron accelerator could be added, further increasing the combinatorial richness of the laboratory.

It may be possible to defend the thesis that the world needs two synchrotrons as similar as the Fermilab main ring and the SPS, or two storage rings as similar as PEP and PETRA. It is certainly true that none of those instruments will sit idle for lack of worthy experiments to carry out. For the largest accelerators now contemplated, however, any possible benefits of duplication become moot. The worldwide budget for high-energy physics is not likely to support more than one proton synchrotron with a beam energy of 20 or 30 TeV. If such a machine is ever to be built, it cannot be as a national project, nor is it likely to be the undertaking of a regional consortium such as CERN. Worldwide sponsorship would be appropriate to such an endeavor; it might be necessary, and a dream of many physicists is that it would be possible.

With a curious blend of pragmatism and idealism physicists have long advocated and practiced international collaboration. CERN is an exemplary model, but there are other robust international endeavors, including some with a broader constituency. In 1961, as a result of the Atoms for Peace movement, an exchange of particle physicists was worked out between the U.S. and the U.S.S.R. for the discussion of undertakings in the mutual interest. The possibility of a "World Accelerator," to be supported by the pooled resources of many nations, was one of the topics considered. Those talks came to an end with the episode of the U-2 flight over the U.S.S.R., but the idea did not die.

Enthusiasm for the World Accelerator surfaced again in 1975, at an international meeting convened in New Orleans by Victor F. Weisskopf of the Massachusetts Institute of Technology. What was strong sentiment at the New Orleans meeting has now been embodied in an official agency, the International Committee for Future Accelerators, which operates under the aegis of the International Union of Pure and Applied Physics. The committee is charged with coordinating the plans of the various national and regional laboratories and with encouraging collaboration among them. Many physicists think such joint ventures are already working well, and are suspicious of official meddling.

A second function of the committee has been received more enthusiastically: it is to provide a forum for the discussion of what is now called the VBA, for Very Big Accelerator. To that end two international workshops have been held, one at Fermilab in 1978 and the second at CERN a few months ago. Both meetings have been concerned largely with establishing the possibilities and limits of accelerator and particle-detector technology, but of course the discussions were held in the context of the physical phenomena to be expected at VBA energies.

It appears that building a 20-TeV proton synchrotron is feasible and would require no major departures from present practice. A large share of the increase in energy would be obtained by the simplest of strategies: increasing the radius of the ring. Very-high-energy colliding-beam devices, on the other hand, might benefit from new techniques. For electron-positron collisions, for example, energies beyond the LEP range might be reached most efficiently by a pair of linacs aimed muzzle to muzzle, or by a combination of linacs and electron synchrotrons. A means for improving the luminosity of colliding beams without accelerating more particles has been suggested; it consists in raising the particle density of the beams by squeezing them down to a microscopic cross section at the point of interaction. That presents a new phenomenon that was recognized by the first workshop sponsored by the International Committee for Future Accelerators: in a small beam the magnetic field becomes large, so that synchrotron radiation in the magnetic field itself becomes an important energy loss. This phenomenon is known as "beamstrahlung," a pun on bremsstrahlung, or braking radiation, the process in which charged particles emit electromagnetic radiation on being decelerated. Burton Richter and his colleagues at SLAC are seriously proposing that colliding 50-GeV linac beams could be a Z^0 factory — a challenge to the Cornell plans.

One thing that most interested physicists agree on is that a world effort should not be resorted to until it is clearly recognized as the only way to achieve the construction of a particular accelerator. It is significant that when serious discussion of a World Accelerator began in 1960, it was a 100-GeV proton synchrotron that was being considered. Now, because of technological advances, 10 times that energy can be achieved by a national effort alone. At the time the International Committee for Future Accelerators was organized a proton synchrotron of 10 TeV and an electron-positron storage ring with colliding beams of 100 GeV seemed appropriate projects for a worldwide effort. The proposed construction of LEP at CERN has preempted the latter possibility. The workshops sponsored by the committee arrived at the conclusion that it would be possible to build electron storage rings with a circumference of about 100 kilometers and about twice the energy of LEP. Above that energy colliding linac beams with energies of up to 350 GeV appear to be more feasible. For protons a synchrotron with a circumference of between 50 and 100 kilometers, with energies in the TeV range and with all the options of antiproton-proton collisions, would also be a viable candidate for the Very Big Accelerator.

If it should become necessary to build such large accelerators in order to study the inner structure of leptons and quarks, and it probably will, then the International Committee for Future Accelerators is providing the necessary international foundation on which such an effort can be built. One is reminded of a fanciful suggestion made many years ago by Enrico Fermi: to build an accelerator that would encircle the world. By that standard an accelerator such as the VBA would not be very big, but the common effort needed to build such a machine might nonetheless help to bind the world together.

The Superconducting Super Collider

An accelerator 20 times as powerful as any now operating could be built by 1995. It would probe matter in unprecedented detail and re-create conditions prevailing near the beginning of time.

. . .

J. David Jackson, Maury Tigner and Stanley Wojcicki
March, 1986

The year is 1995. A pastoral landscape of farmland or prairie gives almost no hint that a tunnel, large enough to walk through and curved into a ring some 52 miles around, lies buried below the surface. Inside the tunnel there is a small tramway for maintaining two cryogenic pipelines, each about two feet in diameter. Within each pipeline is a much smaller, evacuated tube that carries a beam of protons, which are kept on course by powerful superconducting magnets surrounding the tube. With every circuit of the ring the energy of the protons in the two beam pipes is boosted by a pulse of radio waves; in 15 minutes the protons are accelerated around the ring in opposite directions more than three million times.

Suddenly electromagnetic gates are opened and the beam paths are made to cross. Pairs of protons collide, and some of the energy of the collision can be transferred at a rate that far exceeds the instantaneous output of all the power plants on the earth into a region whose diameter is 100,000 times smaller than the diameter of a proton. There, for a time so brief that it is to the second what the second is to 100,000 times the age of the universe, we shall have a glimpse of the universe at the moment of creation. The energy that will be concentrated in

that region at that instant is now found only in the rarest of cosmic rays, but such energy concentration was the prevailing state of the universe in the first 10^{-16} second after the big bang. New elementary particles that could materialize from the energy may show how to explain the origin of mass.

It is remarkable that such a vision is well within the reach of 20th-century technology. The tunnel, the pipeline, the attendant operating systems and an initial complement of particle detectors and computers can be built entirely with available technology—albeit on a scale never before attempted—at a cost of about $4 billion in constant 1986 dollars. Indeed, the basic design of the instrument is already being tested at an energy scale of about 1/20: the scale model is the particle accelerator known as the Tevatron, at the Fermi National Accelerator Laboratory (Fermilab). The scaled-up version would make it possible to study energetic processes that are not accessible to any other accelerator now in operation or seriously contemplated anywhere in the world.

The proposed machine builds on the accumulated experience of more than 50 years in designing and constructing particle accelerators. Moreover, since 1982 the community of high-energy physicists has devoted considerable thought and energy to the design of an economically realistic machine that will have the greatest chance of resolving outstanding theoretical questions about the ultimate constituents of matter. Because it is designed as a colliding-beam accelerator and because superconducting magnets play an essential role in minimizing its consumption of power, the instrument we are planning is called the Superconducting Super Collider, or SSC.

ENERGY AND LUMINOSITY

Particle accelerators can set off the most highly energetic reactions one can study under controlled conditions. Because mass and energy are equivalent, the maximum reaction energy fixes the maximum observable mass of the basic material entities that can be created in the laboratory. Hence the design of a particle accelerator determines the limits of direct, experimental knowledge about the fundamental structure of matter. The maximum observable energy and mass in a particle accelerator depend on the energy of the accelerated particle

beams, the way the energy of the beams is released and the intensity of the beams.

Energy can be released most effectively for creating new particles of high mass if two beams of equal and opposite momentum are made to collide. This conclusion follows from the law of conservation of momentum, which is strictly obeyed in all collisions between particles: the total momentum of the products of a reaction must be equal to the total momentum of the reacting particles. If a particle in an energetic beam is made to collide with a particle in a target at rest, the forward momentum of the energetic incoming particle must be conserved. Thus much of the total energy of the two particles is needed for imparting forward momentum to the particles leaving the scene of the collision.

In contrast, if the momentums of two colliding particles are equal in magnitude but opposite in direction, the total momentum of the pair is zero. In principle none of the total energy of the pair is needed for imparting momentum to the reaction products, and so much of that energy becomes available for creating new particles. Our design now

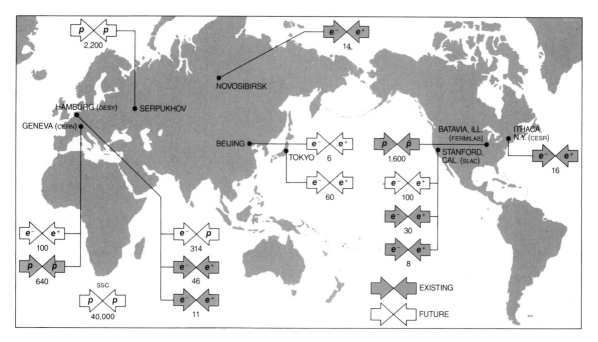

Figure 5.1 COLLIDING-BEAM ACCELERATORS dominate the technology for high-energy experiments in physics. The map shows both existing and planned machines. The accelerators are labeled according to the particles they make collide: e^- for the electron, e^+ for the positron, p for the protron and $\bar{p}$ for the antiproton. The numbers give the maximum total collision energy in billions of electron volts.

calls for a machine in which two counterrotating beams of protons are each accelerated to an energy of 20 trillion electron volts (TeV) and made to collide; the energy of a collision would be 40 TeV, or more than 20 times the collision energy at the Tevatron collider. The energy available for creating new matter at the SSC will be 200 times the energy that would be available if only one such beam were directed at a fixed target.

In a colliding-beam accelerator a good measure of the useful beam intensity is the luminosity of the collider. Because the beams are made up of evenly spaced trains of equal bunches of particles, the luminosity is the number of particles per bunch in one beam multiplied by the number of intercepted particles per unit area in the second beam, all multiplied by the frequency with which the bunches collide. Luminosity is thus an index of the average number of events per second for any given reaction. A high luminosity is needed to guarantee that extremely rare high-energy reactions are generated often enough for physicists to confirm their exis-

Figure 5.2 SIMULATED COLLISION of two protons, each accelerated to an energy of 20 TeV, gives rise to the shower of particles shown in a computer-generated image. Such energies could be reached by the SSC. A 300 GeV Higgs boson materializes out of the energy liberated at the collision site and decays into a W^+ and a W^- boson. The W^+ boson decays into a positron (*solid blue line*) and a neutrino (*broken yellow line center to upper left*), and the W^- boson into two light quarks, each giving rise to a shower of composite particles made up mostly of pions (*solid green lines*). Other collision by-products are photons (*red*), muons (*white*) and baryons (*purple*). (Simulation by J. Freeman of Fermilab, using the ISAJET model devised by F. Paige at Brookhaven.)

tence. The current design of the SSC calls for a luminosity of 10^{33} in standard units, about 1,000 times that projected for the Tevatron.

Throughout the development of this design we have been acutely aware of the need for economic optimization; indeed, the SSC was recommended in 1983 by the High Energy Physics Advisory Panel to the U.S. Department of Energy only after an agonizing decision to abandon several other funding requests for particle accelerators. In response to the recommendation the Energy Department commissioned a research-and-development program, whose purpose is to prepare a technically realistic and economically optimum design and to give a detailed cost estimate. The program, coordinated by the SSC Central Design Group of the Universities Research Association, is being carried out at the Brookhaven National Laboratory, Fermilab, the Lawrence Berkeley Laboratory and the Texas Accelerator Center, in universities and in industry, with valuable assistance from abroad. We and our colleagues shall submit the report of that program next month, and it will serve as one important factor in a

decision on further support of the SSC by the Energy Department.

Continued engineering development of the SSC will further lower its cost. Strive as we will, however, the final cost will be high compared with the costs of other scientific instruments. That cost is driven primarily by one of the basic laws of nature: in order to probe the structure of matter at increasingly fine resolution one must accelerate particles to increasingly high collision energies. To justify the costs of such a probe we shall try to explain why particle physicists think the extreme physical conditions available at the SSC are important to explore and how such conditions are made possible by the current design.

THE STANDARD MODEL

In the past two decades remarkable progress has been made in identifying the basic constituents of matter and the fundamental forces by which they interact. According to what is now called the standard model of elementary processes, all matter is

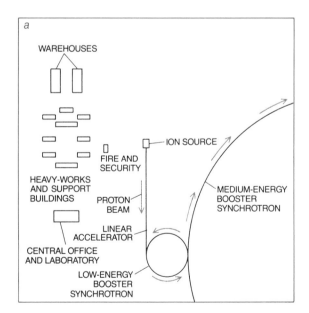

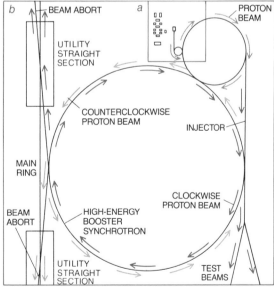

Figure 5.3 SCHEMATIC MAPS OF THE SSC portray the machine at two scales: the central office and laboratory complex (*a*) and the cascading series of three synchrotrons that preaccelerate the protons (*b*). Hydrogen atoms are ionized and the protons that make up their nuclei are accelerated (*purple*) by a linear accelerator to an energy of .6 GeV. The protons then enter the synchrotrons and are acceler-

ated to an energy of 1 TeV in three stages; the beam is sent alternately clockwise and counterclockwise (*red, blue*) around the high-energy booster synchrotron for injection into the main collider rings. In a main ring each beam is further accelerated to 20 TeV. The protons are then made to collide and the collision by-products are monitored (see inset in Figure 5.4 between Queens and Brooklyn).

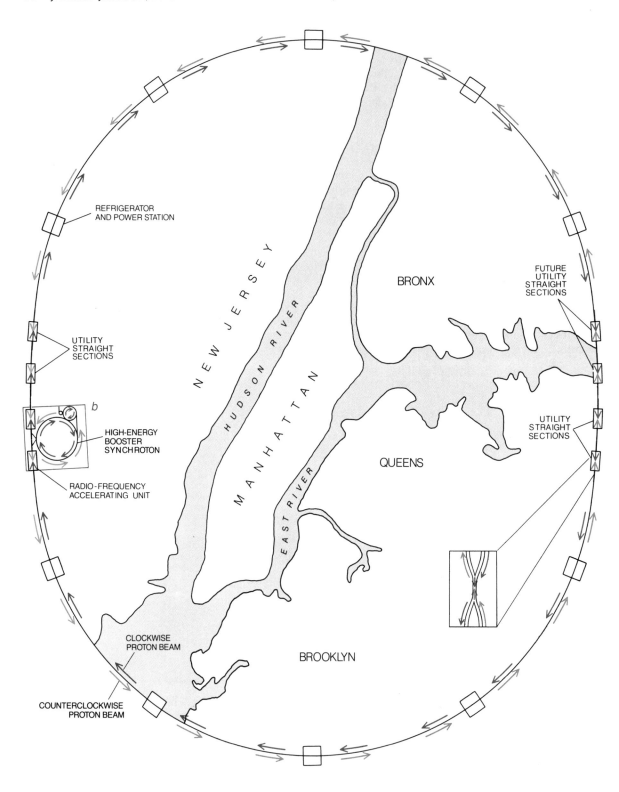

REFRIGERATOR
AND POWER STATION

BRONX

FUTURE
UTILITY
STRAIGHT
SECTIONS

UTILITY
STRAIGHT
SECTIONS

N E W   J E R S E Y

H U D S O N   R I V E R

M A N H A T T A N

E A S T   R I V E R

b

HIGH-ENERGY
BOOSTER
SYNCHROTON

QUEENS

UTILITY
STRAIGHT
SECTIONS

RADIO-FREQUENCY
ACCELERATING UNIT

CLOCKWISE
PROTON BEAM

BROOKLYN

COUNTERCLOCKWISE
PROTON BEAM

Figure 5.4 THE MAIN RING OF THE SSC superimposed on an outline of New York City drawn to the same scale. Superconducting magnets are cooled by liquid helium surrounded by liquid nitrogen, both of which are pumped outward along the main ring from 10 refrigerator and power stations.

made up of quarks and leptons, whose interactions with one another are mediated by the exchange of so-called gauge particles. It is also thought there are four basic kinds of interaction: electromagnetic, weak, strong and gravitational.

For example, the electron is classified as a lepton, and its electromagnetic interactions with the proton are mediated by a gauge particle called the photon. Beta decay, which is central to nuclear burning in the sun, is a result of the weak interaction, and it is mediated by the exchange of the gauge particles called weak vector bosons. The proton, the neutron and many other particles are classified as hadrons, and they are made up of three fractionally charged quarks. The quarks are held together by a strong interaction called the color interaction, and that interaction is mediated by the exchange of eight kinds of gauge particles called gluons. By analogy with these three interactions it is assumed that another gauge particle, the graviton, mediates the gravitational interaction, but such a particle has not been detected. In all it is now believed there are at least six quarks and their six corresponding antiquarks, each in three varieties of "color," six leptons and their six corresponding antiparticles, one photon, three weak vector bosons, eight gluons and perhaps a graviton.

The standard model is based mainly on data from the large proton synchrotrons at Fermilab and at CERN, the European laboratory for particle physics, from the Stanford Linear Accelerator Center (SLAC), from the electron-positron colliders at Cornell, Hamburg and Stanford and, recently, from the proton-antiproton collider at CERN. Delicate low-energy experiments have also made important contributions to understanding. The quarks were originally introduced purely as theoretical entities after hundreds of hadrons had been discovered, in order to restore some underlying organization to the proliferating "elementary" particles. Quarks acquired a certain shadowy reality from the results of a large variety of experiments; it was not until 1974, however, that belief in their existence was firmly established by the simultaneous discovery at SLAC and at

Brookhaven of the J/psi particle, which had been predicted on the basis of the quark hypothesis.

A central component of the standard model is the electroweak theory. In the present versions of that theory the six quarks and the six leptons are grouped into three generations; a pair of quarks and a pair of leptons are assigned to each generation. The electromagnetic and the weak interactions are described as different aspects of one underlying interaction called electroweak. The electroweak theory makes precise predictions about a great variety of phenomena, and it has been confirmed in detail by many experiments. Its most spectacular confirmation came at CERN in 1983 with the detection of the three weak vector bosons, the W^+, W^- and Z^0 particles.

The electroweak theory maintains a tradition that has characterized scientific thinking since its origins in ancient Greece: the unification of diverse phenomena under a single set of concepts. Indeed, to many physicists it is the paradigm of the kind of theory that may one day succeed in giving a unified account of all four fundamental interactions in nature. According to the electroweak theory, the unification of the weak and the electromagnetic interactions is manifest only at extremely high energies. At such energies the interactions are equivalent because the masses of the gauge bosons that mediate the two interactions are effectively zero. The full symmetry of the two interactions can come into play without inhibition.

The hypothesis of such a symmetry at high energies contrasts sharply with the properties of the two interactions in the ordinary laboratory environment. There the range of the weak interaction is roughly 1,000 times smaller than the diameter of the atomic nucleus, whereas the range of the electromagnetic interaction is infinite. According to the electroweak theory, this difference is a consequence of the fact that the weak gauge bosons are very heavy particles, whereas the mass of the electromagnetic gauge boson (the photon) is zero. The symmetry of the two interactions is said to break.

THE ORIGIN OF MASS

Why is the symmetry of the weak and the electromagnetic interactions so badly broken at ordinary energies? The question becomes even more compelling because in the theoretical formulation of electroweak symmetry both the photon and the weak vector bosons initially have zero mass. Thus the observed heavy masses of the weak vector bosons arise out of the electroweak symmetry breaking, and the above question becomes in effect: What is the origin of mass?

The origin of mass is a problem central to determining the fundamental capabilities of the SSC. The analysis of electroweak symmetry breaking leads to a variety of potential theoretical scenarios, but they all share a common property: the evidence for or against any of them must become manifest at the collision energies now proposed for the SSC. There are, to be sure, many other questions arising from issues both within the standard model and beyond its scope that will be addressed by the SSC, and there are almost surely entirely unknown physical processes whose discovery the SSC will make possible. Nevertheless, most of the guidance we are able to get from theory as to the proper scale of the SSC comes primarily from considering the problem of the origin of mass.

In the simplest version of electroweak dynamics the spontaneous symmetry breaking arises out of an electrically neutral field called the Higgs field, named after Peter W. Higgs of the University of Edinburgh. The Higgs field, if it exists, must assume a uniform, nonzero background value even in the vacuum. The idea that the vacuum "contains" anything, even a uniform, nonzero field, runs counter to the popular notion of the vacuum as empty space. In quantum mechanics, however, the air of paradox about such a result has been dispelled for some time. The quantum-mechanical vacuum constantly fluctuates with activity, whether or not the Higgs field is real.

The interaction of a particle with the Higgs field contributes to the energy of the particle with respect to the vacuum. That energy is equivalent to a mass. In the simplest model of the Higgs field the masses of the quarks, the leptons and the weak vector bosons are all explained as a result of the interaction with a single Higgs field. There is always a particle associated with a quantum-mechanical field, and so in the simplest form of the Higgs mechanism for symmetry breaking there is one Higgs particle associated with one Higgs field. If the Higgs particle exists, it should be possible to detect it, but searches so far have turned up nothing.

THE HIGGS PARTICLE

One problem is that the mass of the Higgs particle is virtually unconstrained by theory. It could plausibly be as light as a few billion electron volts (GeV) or as

heavy as 1 TeV. If the mass is less than 50 GeV, it should be found at the electron-positron colliders expected to be built and running by the end of this decade, such as the Stanford Linear Collider (SLC) or the Large Electron-Positron Collider (LEP) at CERN. If its mass is between 50 and 200 GeV, the Tevatron collider at Fermilab should be able to produce it, although extracting firm evidence for it amidst a welter of other particles in that mass range may be experimentally difficult.

For Higgs masses much greater than 200 GeV it becomes less appropriate to describe the Higgs as an elementary particle. Quantum mechanics teaches that the shorter the lifetime of a particle is, the less certainty there is about its energy, or equivalently its mass. If the Higgs mass is greater than 200 GeV, it decays into two W or two Z particles, and its lifetime must be so short that its mass is effectively "smeared out" over a broad range. One is left to ponder what sense there is in regarding as a particle an entity that lacks definite mass.

If the mass of the Higgs is as great as 1 TeV, the electroweak theory predicts that entirely new phenomena must emerge at energies of 1 TeV or more. In such circumstances the electroweak interaction becomes strong. Since electroweak dynamics governs the interactions of leptons and quarks, such particles may combine at energies of 1 TeV or more into composite particles with startling new features. In any case, if one is to confirm or disprove the Higgs mechanisms throughout the potential mass range of a recognizable Higgs particle, the SSC will be needed.

Many physicists think the simplest form of the Higgs mechanism for symmetry breaking is only a low-energy approximation to reality. One reason stems from the fact that the Higgs particle, if it exists, cannot have a spin. If the Higgs particle had a spin, the Higgs field would have spin as well, and the mass of an ordinary particle would depend on its orientation in the vacuum. No such rotational dependence has ever been observed.

In quantum mechanics the spin of a particle can take on only discrete values, and particles that have integral spins (0, 1, 2 and so on) are sharply distinguished from particles that have odd-half integral spins (1/2, 3/2 and so on). Particles that have integral spin are called bosons, and so the spin-0 Higgs particle is a boson, just like the observed spin-1 gauge bosons such as the photon that mediate fundamental interactions. Particles having odd-half integral spin are called fermions, and they include all the quarks and leptons.

To calculate the mass of the Higgs boson one must make certain assumptions about physical processes at high energies. If the Higgs boson is an elementary particle, its calculated mass varies widely even for small changes in such assumptions. Such mathematical sensitivity has no natural physical interpretation; furthermore, it is not a characteristic of the expressions for the masses of spin-1/2 particles. Hence some theorists have proposed that one can avoid the mathematical difficulty and retain the necessary spin-0 Higgs boson if the boson is a composite particle instead of an elementary one, made up of two spin-1/2 fermions. Such composites are common in other domains of physics: the spin-0 pion, for example, is a composite particle made up of two spin-1/2 quarks. The spins cancel to give a composite without spin because the two quarks spin in opposite directions.

A composite Higgs boson would require the existence of an entire new family of heavy, spin-1/2 particles called techniquarks. Such particles would be subject to a new strong interaction called the technicolor interaction, understood by analogy with the strong, color interaction that binds quarks into hadrons. Not only would techniquarks bind to form the Higgs boson but also they would bind to form a plethora of other composite techniparticles such as technipions, technivector mesons and so on. Such new particles would be quite heavy, but at least some of them must have masses of roughly between 50 and 500 GeV (see Figure 5.5). At accelerators such as the Tevatron the number of such particles generated would be small, and they would be hard to detect against the background even if their masses are near the low end of the expected range. To test the theory one must employ a collider with a beam energy higher than several TeV, such as the SSC.

SUPERSYMMETRY

Another theoretical program with many attractive features is called supersymmetry, and it could provide an alternative to the simple Higgs mechanism for explaining the origin of mass. In a supersymmetric world every particle, including the Higgs boson, has a partner identical in every way except in its spin. To every ordinary fermion there corresponds a supersymmetric, spinless boson; for example, the spin-1/2 electron and quark have the spin-0 partners selectron and squark respectively. To every ordinary boson there corresponds a supersymmetric, spin-1/2 fermion; for example, the supersym-

metric partner of the spin-1 photon is the spin-1/2 photino, the partner of the spin-1 gluon is the spin-1/2 gluino and the partner of the spin-0 Higgs boson is the spin-1/2 Higgsino.

If the supersymmetric particles existed in nature as exact copies of their counterparts except for spin, most of the supersymmetric particles would by now have been seen in abundance. Many searches have been made, however, and no evidence for the supersymmetric partners has been found. One might therefore suppose the appeal of supersymmetry would be on the wane, but that appeal persists for a number of reasons. One is that the existence of the supersymmetric partners would solve the problem of mathematical sensitivities in the theoretical expression for the mass of the Higgs boson. A second reason not to abandon supersymmetry is that it may be a broken symmetry in our world, just as electroweak symmetry is. A broken supersymmetry might give rise to supersymmetric particles that are substantially heavier than their ordinary partners.

No one clearly understands how sensitive mass is to symmetry breaking. For example, it is already known experimentally that the mass of the selectron, if it exists, must be at least 40,000 times as great as the mass of the electron. Does this ratio imply that supersymmetry must be "badly" broken? No one knows. What is known is that if supersymmetry turns out to be correct, it too, like the technicolor theory, introduces a new world of particles. Most of them must be quite massive; if they were not, they would already have been detected. Powerful new accelerators will undoubtedly be needed to find them.

There is a third and more general reason not to abandon supersymmetry, or for that matter any other theory such as technicolor that has a chance of explaining the mysteries of electroweak symmetry breaking and the origin of mass. No matter whether the Higgs boson is composite or elementary, whether or not it is embedded in a supersymmetric family of particles or indeed whether or not it exists at all, a general quantum-mechanical principle guarantees that new physical phenomena, deeply related to the origin of mass, should begin to emerge at energies of about 1 TeV. These phenomena must arise because if the existing standard model is extrapolated without any corrections into that energy domain, the probabilities calculated by the theory for certain interactions become greater than 1. Because no real probability can be greater than 1, the theory as it stands cannot be complete.

Since the correct theoretical extension of the stan-

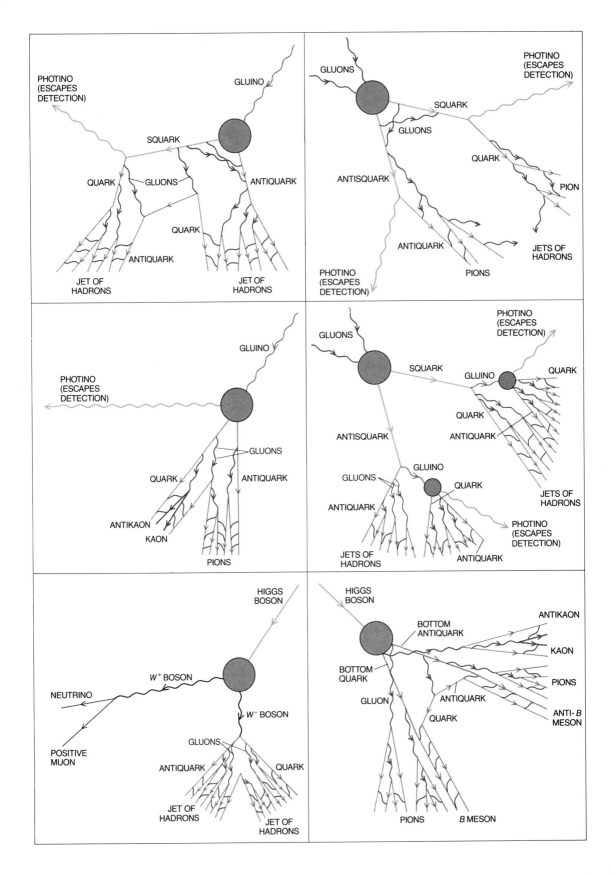

Figure 5.5 SIX HYPOTHETICAL INTERACTIONS at the SSC. The gluino and the squark are the so-called supersymmetric partners of the gluon and the quark. If both particles exist and the mass of the gluino is greater than the mass of the squark, the particles could decay as is shown in the top two diagrams. If the mass of the squark is the greater, the particles could decay as is shown in the middle two diagrams. Gluino decay is shown at the left. Two kinds of decay of the Higgs boson are shown in the diagrams at the bottom. The process at bottom right would predominate if the mass of the Higgs boson were about 50 GeV; the one at bottom left would be typical if the mass were 200 GeV or more.

dard model to very high energies is not known, the exact nature of the new physical phenomena cannot yet be described. If the Higgs boson is quite massive, one possibility already mentioned is that the electroweak interaction becomes a strong one. On the other hand, if the Higgs boson turns out to be light, its small mass could well be explained by supersymmetry. In that case the energy domain of roughly a few TeV would abound with the supersymmetric partners of known particles. The ability to probe that energy domain is therefore an extremely important goal for the basic understanding of matter.

THE 20 TeV REQUIREMENT

Theories such as technicolor and supersymmetry yield specific predictions for the discovery limit of a collider that has a given energy and luminosity, or in other words the largest mass a hypothetical particle can have if it is to be created and detected at the collider. One might think a machine that brings about collisions of protons with a total energy of 40 TeV would make available roughly that amount of energy for the creation of new particles. Unfortunately only a fraction of the collision energy is actually released. All hadrons such as protons and antiprotons are composite systems, each one rather like a sack of marbles. The total energy is divided among the quarks, antiquarks and gluons that make up each hadron, and a collision can release only roughly the amount of energy carried by any two colliding constituents. For example, the Tevatron, with a total energy of 1.8 TeV, can thoroughly explore a range of masses up to only about .3 TeV.

To give some sense of how the largest detectable mass varies with collider design, consider the possible existence of the gluino and the squark. For a proton-antiproton collider whose beams are each accelerated to an energy of 6 TeV and whose luminosity is 3×10^{31} in standard units, the heaviest detectable gluino or squark would have a mass of about .4 TeV. With existing technology such a collider could be constructed in the LEP tunnel at CERN. At the SSC, where the luminosity would be increased over such a LEP collider by a factor of 30 and the energy increased by a factor of three, the heaviest detectable gluino or squark would have a mass of about 1.5 TeV. Similarly, the SSC could detect Higgs bosons as heavy as 1 TeV, new quarks as heavy as 2 TeV and new gauge particles as heavy as 6 TeV. Figure 5.6 illustrates the discovery limits for these cases. Here a discovery is defined as the creation of 10 or more uniquely identified events in one year of data taking.

The numerous theoretical estimates and machine assumptions can be challenged in detail, but the message is clear. To probe effectively into the range of particle masses on the order of 1 TeV, a collider with the design characteristics of the proposed SSC is prudent. Less energy or less luminosity begins to compromise the discovery potential. Moreover, if the maximum available energy lies just below the threshold for the onset of some radically new physics, that physics will go undiscovered no matter what the luminosity is. The need for the highest feasible beam energies is paramount.

PROTON PATH

In order to understand how the SSC will achieve its design characteristics, it will be helpful to trace the path of the protons through the collider from their source to the sites of their collisions (see Figures 5.3 and 5.4). The protons begin their journey as the nuclei of ionized hydrogen atoms in a gas. They are extracted from the gas by suitably arranged electrodes and emerge with a kinetic energy of a few thousand electron volts.

From there they enter a linear accelerator, a sequence of electrodes that accelerates the protons in a series of small pushes. In effect the protons are carried along by a precisely timed wave of potential difference that moves across the electrodes. The acceleration elevates the beam energy to 600 million electron volts. The beam then enters the first in a cascade of four synchrotron rings.

In synchrotron acceleration a uniform magnetic field forces the protons to follow a predetermined path. In a special section of the synchrotron ring the

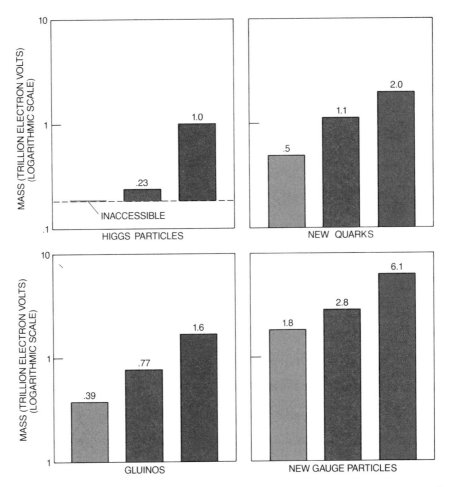

Figure 5.6 DISCOVERY LIMIT is the largest mass a new, hypothetical particle can have if it is to be detected by a colliding-beam accelerator of a given beam energy and luminosity. The graphs show how the discovery limits depend on accelerator beam energy and luminosity for four kinds of hypothetical particles. The graphs also show the lower limit for the mass of the Higgs boson: .18 TeV. The light-color graphs give the discovery limit for a proton-antiproton collider whose beam energy is 6 TeV and whose luminosity is 3×10^{31}. The light gray graphs give the limits for a 20-TeV proton-antiproton collider whose luminosity is also 3×10^{31}. The dark-color graphs show the limits for the SSC: a 20-TeV proton-proton collider whose luminosity is 10^{33}.

beam passes through a linear accelerator, and its energy is thereby increased with each circuit of the ring. As the energy increases, the strength of the magnetic field is also increased in order to keep the protons in their closed orbit. The frequency of the traveling wave that accelerates the protons within the linear-accelerator section must be synchronous with the frequency at which the protons circle the ring; the frequency is in the radio region of the electromagnetic spectrum. Modern acceleration systems are so efficient that the accelerator section

will occupy a length of only 75 feet along the entire 52-mile circumference of the SSC.

The cost of the magnets that steer the beam is proportional to the diameter of the region in the beam pipe in which a uniform magnetic field is needed. The necessary field diameter depends on the diameter of the beam, and the beam diameter depends on, among other things, the ratio of the momentum of the particles in a direction transverse to the beam to their momentum along the beam path. The ratio decreases continuously because the acceleration system increases the momentum of the particles only along the beam path. In a large synchrotron the decrease in the beam diameter is exploited to minimize the diameter of the uniform magnetic field needed at each stage in the acceleration and thus minimize the overall cost. In our current design a cascade of synchrotrons sends the protons through a series of progressively narrower beam pipes. The first synchrotron will accelerate the protons from an energy of .6 GeV to about 8 GeV, the second will boost them from 8 to about 100 GeV and the third will boost them from 100 GeV to 1 TeV. The main, large synchrotron ring of the SSC will accelerate them to their final energy of 20 TeV.

At full energy the diameter of the beam will be a fifth of a millimeter. To enhance the luminosity of the collider the beams will then be focused at the collision points by powerful magnetic lenses into even tighter bunches. After focusing the bunches will be slim cylinders about 10 micrometers in diameter and 15 centimeters long. Each bunch will carry about 10 billion protons, and so the density of the bunch will be roughly one ten-thousandth the den-sity of the molecules in the air at ordinary temperature and pressure. Because the probability of a collision between the protons in two counterrotating bunches is small, the bunches can interpenetrate repeatedly at the interaction points for many hours without need of replenishment.

SUPERCONDUCTING MAGNETS

The need to bend and focus the proton beam makes the magnet system a key element of any synchrotron, and in the past two years a considerable national effort has been devoted to studies of different kinds of magnets that might be adopted at the SSC. In principle it would be possible to build the SSC with copper-conductor electromagnets such as the ones used in the synchrotrons of the 1950's and 1960's. Such a ring, built to accelerate protons to an energy of 20 TeV, would consume at least four billion watts of power and lead to impractically high operating costs. Moreover, the magnetic properties of iron and the capacity of copper to carry electric current limit ordinary electromagnetic fields to a

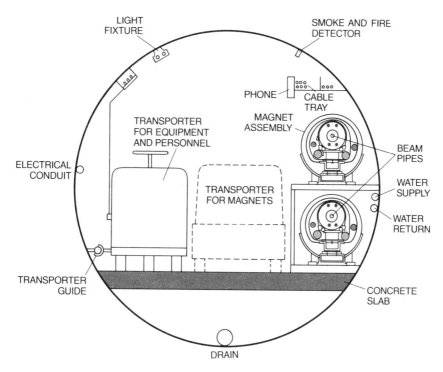

Figure 5.7 MAIN TUNNEL proposed for the SSC is shown in cross section. The two beam pipes for the counterrotat- ing beams of protons are at the right. The tunnel is about 10 feet across.

strength of about 2 teslas, or 20,000 gauss, which is roughly equivalent to the fields generated in the electric motors of home appliances.

Superconducting magnets can minimize both problems. They can sharply reduce the total power consumption of the acceleration system, and they can create magnetic fields several times stronger than conventional magnets can. A stronger magnetic field makes it possible to confine the protons to a tighter orbit for a given energy, and so it reduces the required size of the synchrotron ring.

In the conventional electromagnet wound with copper wire power is dissipated in forcing electric current through the wire. When certain metals and metal alloys and compounds are cooled to below a certain critical temperature, however, they become superconducting: they carry electric current without resistance. In a magnet wound with superconduct-

ing wire the only power required is that needed to maintain the wire below the critical superconducting temperature (see Figure 5.8). The refrigerators needed for the SSC would consume about 30 million watts of power, which is somewhat less than the total required by the largest accelerator complexes today.

In 1983 the first superconducting synchrotron, the Tevatron, went into service. Its successful operation since then has provided data that have been invaluable for planning the SSC. Last September we and our colleagues in the SSC Central Design Group reached a milestone by making our final choice for the design of the superconducting, bending magnets. The bending magnets will have a field strength of about 6.6 teslas, and that strength will give a circumference of about 52 miles for the main synchrotron ring.

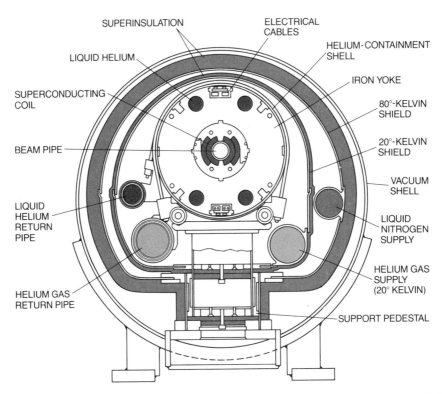

Figure 5.8 DETAIL OF MAGNET ASSEMBLY to be mounted in the tunnel of the SSC in schematic cross section. One of the proton beams passes through an evacuated beam pipe in the central, upper part of the assembly. The pipeline is surrounded by coils of superconducting wire; current passing without resistance through the wire creates the enormous magnetic field needed for bending the proton beam. The rest of the system keeps the magnet at the low temperatures necessary to maintain the superconductivity of the coils. The liquid helium refrigerant is held at 4.35 degrees K. and is surrounded by piping that carries liquid nitrogen at 80 degrees K.

Both the superconducting coil of the magnet and a surrounding yoke of iron will be held at a temperature of 4.35 degrees Kelvin by a pressurized flow of liquid helium. Ten refrigerators will be spaced uniformly around the ring, and each one will supply a coolant stream along the ring out to about four kilometers in both directions. An intermediate heat shield of liquid nitrogen, maintained at 80 degrees K., will sharply reduce the heat energy that is incident on the liquid helium. To stabilize the entire system against temperature changes and mechanical vibrations the ring will be housed in a tunnel. The tunnel will then be covered with at least six meters of earth in order to safely absorb any ionizing radiation that might be generated if the beam were to strike the walls of the beam pipe.

DETECTION APPARATUS

Because the SSC will explore a previously unknown energy regime, no one can predict with certainty the properties of the most interesting events. Hence it is essential to maximize the variety and flexibility of the particle detectors planned for the machine. Experiments with the initial complement of detectors will guide the ultimate experimental program.

The collisions between the counter-rotating beams of protons will be generated by forcing them to intersect head on, or at a small angle, at one of several points along the ring. At such interaction points there will be elaborate detectors housed in experimental halls. The detectors will record the passage of particles emerging from the collisions and sift through them electronically in order to identify the ones of potential interest.

We expect to generate up to 100 million collisions per second in each interaction region of the SSC, and so the electronic sifting is not at all trivial: one must find a needle or at most a few needles in a haystack of data. Furthermore, once such a potentially interesting collision event is identified, the event must be subjected to additional, more detailed tests.If it still passes the tests, enough information must be collected and recorded about the tracks emerging from the collision to allow a later reconstruction of what actually took place. The entire process must be done as quickly as possible to minimize the "dead" time in which the detector and its associated computer systems are not recording collision data.

Since the need to minimize the dead time requires that one record only a small fraction of all the collisions, one must design an electronic trigger for the interesting phenomena. Here one is guided both by past history and by theoretical ideas. For example, the W^+, W^- and Z^0 bosons were signaled by the emission of energetic electrons and muons and, in the case of the two W bosons, by the emission of neutrinos. The Higgs boson is expected to give rise to decay signatures similar to those of the weak bosons. Theory predicts it should decay into the most massive particles possible, such as heavy quarks or W bosons. The quarks or W bosons should in turn give rise to electrons, muons and neutrinos. Neither the Higgs boson nor the heavy quarks and W bosons can live long enough to be directly detected, and so one must infer their presence from their by-products.

Because electrons and muons are electrically charged, their presence can be directly detected; the detection of the neutrinos, however, is subtler. The presence of the neutrino can be inferred by measuring the total momentum, in directions transverse to the proton beams, of the particles emerging from a collision. Since the total momentum of the colliding protons in directions transverse to the proton beams is essentially zero, the total transverse momentum of the emerging particles must also be zero. Thus the total momentum carried by particles in one direction transverse to the beam must be balanced by the total momentum carried in the opposite direction. If it is not, one can assume the imbalance was caused by the passage of a neutrino, a particle that interacts so rarely with matter that it is almost never detected.

The first trigger at the SSC might require only the passage of a moderately energetic electron or muon. Higher levels of decision making might then integrate increasingly sophisticated information from several kinds of detector about the energy and flight path of the emerging particles. For some kinds of triggering it may be sufficient to build a detector that surrounds only part of the collision site. In most cases, however, general-purpose detectors are needed that surround the collision site as completely as possible. We can best illustrate the functions of a detector by describing the main features of the general-purpose detector.

GENERAL-PURPOSE DETECTORS

For a great many measurements a general-purpose detector with a magnetic field in its central region gives a decided advantage, but there is a price attached. The increased complexity of the tracking and the inert material needed for the magnetic coil make measurement hard, and the coil adds significantly to the cost. Hence an optimum design for the initial detectors at the SSC might include one detector with a magnetic field and another detector without one.

The components of such detectors are quite similar (see Figure 5.9). The innermost component will probably be a so-called vertex detector, designed to measure the tracks of particles as precisely as possible near the vertex, or site, of the proton collision. The particles then pass outward into a central tracking chamber that measures the direction and, if the magnetic field is present, the curvature of the tracks of particles that carry electric charge. Just outside the tracking chamber, or perhaps interspersed with it, there may be detectors that identify the electrons.

The next group of active detectors are the calorimeters, which measure the total energy of all the particles emitted in directions greater than some critical angle to the direction of the incident proton beams. One purpose of the calorimeters is to infer the presence of high-energy neutrinos that result from "hard," or head-on, collisions between the constituents of two protons. Remember that in principle the neutrinos can be detected by finding an apparent violation of the conservation of transverse momentum. For particles moving close to the speed of light, momentum is equivalent to energy. In practice, therefore, one can detect an imbalance in the transverse momentum of the collision by-products by detecting an imbalance in their transverse energy.

The measurement of transverse energy could be done by individually measuring the energy of each

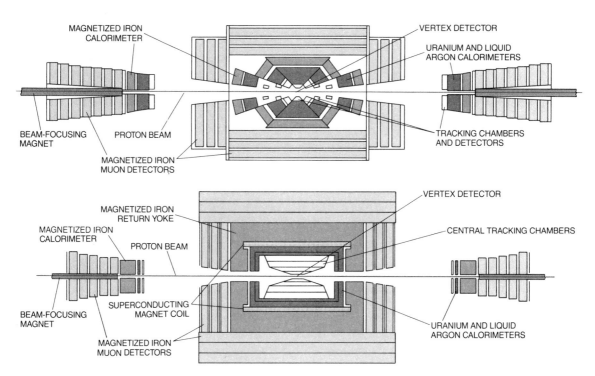

Figure 5.9 GENERAL-PURPOSE PARTICLE DETEC-TORS at the SSC. The upper detector has no magnetic field around the central tracking chambers; the lower one incorporates a superconducting magnet that generates a field along the direction of the beam. Both detectors have a forward part designed to detect the particles moving almost parallel to the beam. An inner part is designed to measure various properties of each charged track. Such properties can later be used to reconstruct the path of a particle. The outer part of each detector is a calorimeter, which measures the total energy deposited in each of its segments. The outermost subsystems identify the muons and measures their energies. All other known particles except the neutrinos are completely absorbed by the calorimeter.

particle coming out of the collision. Such a measurement, however, would be prohibitively difficult. Each collision at the SSC could easily generate more than 100 particles, and many of the particles are expected to emerge in jets, or tight bunches, making them even harder to distinguish. Furthermore, a standard technique for determining the energy of a charged particle, which depends on measuring the curvature of its path in a magnetic field, cannot be applied to a neutral particle.

The calorimeters measure the total absorption of energy in some medium without distinguishing the separate contributions of individual particles. In spite of this limitation the direction of the energy deposits can be determined by segmenting the calorimeters. Thus the total transverse energy and the direction of jets can be measured quite well, and from the data one can determine whether or not a neutrino was generated by the collision.

The calorimeter in the general-purpose detector doubles as the first layer of the system for identifying muons, which are often important signals of interesting events. Outside it there will be several layers of magnetized iron, with tracking chambers interspersed among the layers. The aim in the design here is to allow for redundancy in the measurement of the momentum of the muons: relatively unimportant decays can readily simulate the passage of an energetic muon.

FRONTIERS OF HIGH-ENERGY PHYSICS

We have stressed that the creation and detection of particles having masses above 1 TeV at the SSC will extend knowledge about elementary processes well beyond the compass of the standard model. In particular it will address the fundamental problem of the origin of mass and the problem of symmetry breaking in the electroweak theory. The SSC will also make fundamental contributions, however, to many other open questions. So far, for example, there appear to be three generations of quarks and leptons. Are there more? Why do the quarks and leptons have progressively greater masses in successive generations? Are the quarks always bound to-

gether to form hadrons or shall we ultimately see manifestations of free quarks? Are the quarks and leptons related, and if they are, how? Why do weak interactions show a handedness? Are quarks and leptons really elementary entities or are they built up from some more basic constituents? Does quantum mechanics continue to apply at smaller and smaller scales? Can gravity, as well as the color interaction, be treated in a consistent way by quantum mechanics and perhaps unified with the other known forces?

In the past decade there have been several attempts to extend the partial unification found in the electroweak theory to a grand unification of the electromagnetic, weak and color interactions. Even more recently a development called superstring theory has extended the theory of supersymmetry to a mathematical formalism that may one day bring about an even grander unification: the unified understanding of all four fundamental interactions, including gravity. Out of such grand unified theories has emerged a realization that particle physics has something to say about the earliest epochs in the history of the universe and that cosmology has something to say about particle physics.

Astronomers now believe the universe began cataclysmically in the big bang. In the almost unimaginably hot, primordial universe just after the big bang the full symmetry of nature's laws must have been manifest. Both the study of the very large and the study of the very small thus converge on a common point of view: in order to continue probing nature's underlying unity and simplicity one must build instruments that investigate domains of progressively higher energy. The discoveries in such domains cannot be fully anticipated, but experience teaches us it is often the unexpected discovery that triggers a deeper scientific understanding of the world. The SSC, ambitious yet feasible, would take us to domains of energy never before encountered, where the real discoveries can only be guessed at, and it would give us access to the events that took place almost immediately after the beginning of time. The opportunity and the challenge presented by the SSC will excite all who share our desire to understand the natural world.

Postscript:
Section II, Tools

· · ·

Richard A. Carrigan, Jr.

It is remarkable to see how well the course of accelerator development has paralleled Robert Wilson's game plan described in Chapter 4. The construction of a major accelerator is a first class scientific technical achievement, but more impressive is how well those investments in accelerator technology have paid off in scientific advances.

Several of the accelerators discussed have now operated successfully. By converting their Super Proton Synchrotron (SPS) to colliding operation, CERN won the "race" to build a proton-antiproton collider. It was done by building an antiproton source using Simon van der Meer's stochastic cooling scheme as described in Chapter 7 by David B. Cline, Carlo Rubbia and van der Meer. Rubbia's team built the gigantic UA1 detector and used the collider to discover intermediate vector bosons, the famous W and Z particles. Van der Meer and Rubbia received the Nobel Prize for that work in 1984.

Fermilab concentrated on the Tevatron, which produced its first beam in 1983 and is now operating at 900 GeV as the highest energy accelerator in the world. The Tevatron, by far the largest superconducting system ever built, has operated so smoothly that work on several other major new superconducting accelerators is underway. An antiproton source was commissioned and 1.6 TeV head-on collisions of two beams were achieved in 1985 with serious colliding-beam physics starting in 1987. •

In the years since Wilson's article a remarkable panoply of new and powerful accelerators have appeared. The first was an electron-positron colliding-beam machine called TRISTAN, located in Japan; it now provides 60 GeV of energy in the center-of-mass. In 1989 the Stanford Linear Collider (SLC), a so-called single-pass collider, produced Z particles for the first time. This brilliant innovative device uses the Stanford linear accelerator, now upgraded to an energy of 50 GeV, to accelerate both positrons and electrons, which are focused to a truly microscopic point, a few microns across, where they collide head-on. This scheme is different from all the earlier colliding-beam schemes because the two sets of particles pass by each other only once. The Large Electron Positron Machine (LEP) at CERN is a more conventional storage-ring system located in a 27-km.-long tunnel and will initially have individual beam energies of 55 GeV. Plans call for operation in 1989.

Another new complex, HERA, under construction at DESY in Germany, will provide the first electron-proton colliding-beam system. HERA will have a superconducting proton storage ring quite similar to the Fermilab Tevatron but counterrotating electrons in a second ring will collide with these protons. HERA is being built directly under the city of Hamburg. UNK, the large Soviet superconducting proton accelerator mentioned is also under construction. When that machine goes into full operation in the 1990's it could well be the most powerful fixed target accelerator in the world, with a beam energy of 3 TeV.

This complex of machines should maintain world particle physics activities well into the mid-1990's. But what lies beyond? The possibility of the Very Big Accelerator (VBA) mentioned in Chapter 4 offers exciting opportunities that could extend well into the next century. By now, very serious plans for such an accelerator have been developed as the Superconducting Super Collider (SSC). In Chapter 5 J. David Jackson, Maury Tigner and Stanley Wojcicki, the leaders of the design group, review the exciting plans and the physics that will be opened up by the SSC to be built in Texas.

III

WEAK INTERACTIONS

...

Heavy Leptons

The class of elementary particles of matter that includes the electron and the muon has a new member: the tau. It may be the first in a sequence of charged heavy leptons.

. . .

Martin L. Perl and William T. Kirk
March, 1978

The small family of elementary particles known as the leptons (from the Greek for "light ones") can be distinguished from the two other major classes of subatomic matter by a number of criteria, among which mass is perhaps the least important. For example, the leptons differ from the generally heavier hadrons primarily by being insensitive to the strong nuclear force, the dominant short-range force that binds together the particles of the atomic nucleus. The leptons share this immunity to the strong force with the photon, the massless carrier of the electromagnetic force, which forms a one-member class of its own. Unlike the photon, however, both the leptons and the hadrons are capable of interacting by means of the weak nuclear force, the even shorter-range force responsible for the radioactive decay of nuclear particles.

One of the most striking things about the leptons is that there are so few of them. Here again they stand in contrast to the hadrons, which have proliferated notoriously in recent years, reaching the point where they now total several hundred distinct particles, arrayed in various subclasses. The most familiar of the hadrons are the two main constituents of the atomic nucleus, the proton and the neutron; one of the latest additions to this class is the psi, or *J*, particle, discovered in 1974. (Burton Richter of Stanford University and Samuel C. C. Ting of the Massachusetts Institute of Technology shared the 1976 Nobel prize for physics for their independent discovery of this particle.) The lepton family, on the other hand, has for some time been thought to consist of only four particles (together with their corresponding antiparticles): the electron, discovered in the form of cathode rays more than 80 years ago; the muon, first observed in cosmic-ray showers some 40 years ago, and two kinds of neutrino, one associated with the electron (called the electron neutrino) and the other associated with the muon (the muon neutrino). Neutrinos, long suspected on theoretical grounds, were detected for the first time some 20 years ago.

Another sense in which the leptons are special has emerged just recently. A considerable body of evidence points to the conclusion that hadrons are not elementary particles at all but rather are composite structures built up of the simpler constituents termed quarks. Prior to the discovery of the psi particle all the known hadrons could be accounted for by assuming that they represented various combinations of three different kinds of quark (labeled "up," "down" and "strange") and their corresponding antiquarks. The main significance of the discovery of the psi particle was that it provided compelling evidence for the existence of a fourth kind of quark, which had earlier been named the "charmed" quark. According to the revised quark picture, the psi particle is a hadron consisting of a charmed quark and a charmed antiquark.

There is no evidence that the leptons are anything but pointlike objects, and so it seems that they, unlike the hadrons, are truly elementary particles, in the traditional sense of being indivisible. The list of the already known particles that can still be considered elementary in this sense has accordingly become quite short: four kinds of quarks (and their antiquarks), four kinds of leptons (and their antileptons) and the photon (which is its own antiparticle). In this context the search for additional members of the lepton family has acquired new meaning, because such particles, if any exist, would also be counted among the few genuinely fundamental constituents of matter.

During the past few years a group of physicists led by one of us (Perl) has been conducting just such a search at the Stanford Linear Accelerator Center (SLAC), shown in aerial view in Figure 6.1, as part of a larger experimental program carried out jointly by research groups from SLAC and from the University of California's Lawrence Berkeley Laboratory. We have had at our disposal a potentially powerful tool for creating new leptons: the SPEAR electron-positron storage ring, a device in which counterrotating beams of matter (electrons) and antimatter (positrons) can be made to pass through each other, causing occasional high-energy collisions in which the original particles are annihilated and, out of the resulting flash of energy, new particles are created. With this system we have uncovered evidence of the existence of a fifth kind of lepton. The new particle, which is electrically charged (like the electron and the muon), is much heavier than any of the previously known leptons; indeed, it is heavier than some hadrons. We shall relate here the story of the discovery of the new heavy lepton and its antiparticle, which we have named the tau and antitau.

How does one go about finding a new kind of elementary particle? It helps to follow a few guiding principles. First, have a clear idea of what you are looking for; then you will know when you have found it. Second, if the object must be made artificially, as we had to do in this case, then find a

Figure 6.1 ELECTRON-POSITRON STORAGE RING at the SLAC experimental area. The high-energy end of the two-mile linear particle accelerator is located just out of the picture to the right. The SPEAR colliding-beam storage ring is the oval structure near the bottom. Electrons and positrons generated by the accelerator are injected into SPEAR through the two tangential arms. The two buildings athwart the straight sections of the ring house the interaction regions, where the two counterrotating beams are made to pass through each other, causing the matter-antimatter collisions that lead to the creation of new particles. The charged-particle detector shown in Figure 6.3 is housed in the larger of the two interaction buildings. (Photo by Stanford Linear Accelerator Center.)

way to make it in copious quantities. Third, make sure that the new object has some distinguishing characteristics that will tell you when you have it.

To get an idea of what we were looking for we took our lead from the known properties of the electron and the muon. In other words, we decided to look for a particle that has an electric charge of either −1 or +1 and that is acted on by both the electromagnetic force and the weak force but not by the strong force. Two questions followed. First, in our search for a new charged lepton what mass should we look for? Second, bearing in mind that the electron lasts indefinitely but that the muon decays in about two millionths of a second, what should we expect the lifetime of the new particle to be? (See Figure 6.2.)

The question of mass was a difficult one, because there was (and still is) no theory that accounts either for the observed masses of the muon and the electron or for the ratio of their masses (approximately 200 to 1). All that was known as of four years ago was that some experiments carried out at the ADONE electron-positron colliding-beam facility in Italy had set a lower limit of about 1,000 million electron volts (MeV) on the mass of any new charged lepton: an energy equivalent to roughly 10 times the mass of the muon and 2,000 times the mass of the electron. Our group at SLAC, however, had no idea whether the mass of a new charged lepton, if one existed, would be within reach of our experimental equipment, which could detect particles with a mass as high as 3,500 MeV.

The question of the lifetime of the hypothetical particle, it turned out, had a better theoretical answer: less than a hundred-billionth of a second for any charged lepton with a mass greater than 1,000 MeV. The lightest charged lepton, the electron, is

stable simply because there is no lighter charged particle into which it can decay. The muon is unstable because it can decay into an electron. The muon-to-electron decay does not take place in the simplest conceivable way, however, which would be for the muon to change spontaneously into an electron and a photon through the electromagnetic interaction, even though more than enough energy would be released in the process to produce a photon. Instead the muon is observed to decay through the weak interaction into three particles: an electron, an electron antineutrino and a muon neutrino.

Physicists explain this odd behavior by invoking an empirical rule whose basic significance remains unclear. They ascribe to the electron and the muon separate intrinsic properties, which we shall refer to here as "electronlikeness" and "muonlikeness," and they postulate that each of these properties is exhibited by a group of four related particles. Thus the electron, the electron neutrino, the antielectron (or positron) and the electron antineutrino are all said to exhibit the property of electronlikeness (or antielectronlikeness in the case of the antiparticles), whereas the muon, the muon neutrino, the antimuon and the muon antineutrino all exhibit muonlikeness (or antimuonlikeness). The rule then states simply that in any particle interaction or decay process involving leptons the properties of electronlikeness and muonlikeness must be separately conserved.

It follows that the simple electromagnetic decay of a muon into an electron and a photon is not possible because in the process the muonlikeness of the muon would have to change into the electronlikeness of the electron. The more complicated weak decay of the muon into an electron, an electron antineutrino and a muon neutrino does occur, because the muon neutrino preserves the muonlikeness of the muon, whereas the electronlikeness of the electron is exactly canceled by the antielectronlikeness of the electron antineutrino. (In actual practice particle physicists employ the terms electron lepton number and muon lepton number to denote the properties of electronlikeness and muonlikeness, and the rule in question is called the conservation of lepton number; in our view, however, these formal terms tend to obscure the fundamental mystery of the separate unique properties of the electron and the muon.)

In any event we had to decide on some comparable property for the hypothetical charged heavy lepton we were looking for. One possibility was to assign an electron lepton number (or electronlikeness) to the new lepton, thereby allowing it to decay electromagnetically into an electron and a photon. A more intriguing possibility was to assume that the new lepton-antilepton pair came with their own separate lepton number and their own associated neutrino-antineutrino pair. According to this view, the electron and the muon might be just the beginning of a sequence of charged leptons, each with its own unique "likeness." The general name adopted for these highly speculative particles at the time was sequential charged heavy leptons, with the "heavy" indicating that their masses would be greater than that of the electron or the muon. Soon after the first fragmentary evidence for such a particle was obtained, we began to use the symbol U (for "unknown" particle), but now that we have substantial evidence for its existence we call it tau, after the first letter of the Greek word $\tau\rho\iota\tau o\nu$, meaning "third." The name is meant to indicate that the tau is the third charged lepton in the sequence beginning with the electron and the muon.

As we have noted, an effective search for a new elementary particle requires some idea of its properties, a method for producing the particle in sufficient quantities and a means of distinguishing the new particle from the particles already known. Experiments utilizing the SPEAR electron-positron storage ring at SLAC can satisfy both of the last two requirements. This machine, which began operating in 1972, consists of about 100 magnets in a ring-shaped array some 80 meters in diameter. The electron and positron beams for SPEAR, which are generated by the SLAC two-mile linear accelerator, are injected into the storage ring during a "filling time" of anywhere from 10 to 30 minutes. The beams circulate in a vacuum chamber that passes through the ring of magnets, and the deflection and focusing provided by the magnets hold the beams in stable orbits for periods of several hours. The circulating beams do not encounter each other except at two points on opposite sides of the ring, where they are made to pass through two interaction regions. In order to maximize the chance of collisions each beam contains about 100 billion particles in a single tightly packed "bunch" that is only a few centimeters long. Although each bunch makes a complete trip around the ring more than a million times a second, the chance that a particle in one beam will make a direct hit on a particle in the other beam is

PARTICLE	MASS	LIFETIME	ANTIPARTICLE
ELECTRON (e^-)	.51 MeV	STABLE	POSITRON (e^+)
ELECTRON NEUTRINO (ν_e)	0	STABLE	ELECTRON ANTINEUTRINO ($\bar{\nu}_e$)
MUON (μ^-)	106 MeV	2.2×10^{-6} SECOND	ANTIMUON (μ^+)
MUON NEUTRINO (ν_μ)	0	STABLE	MUON ANTINEUTRINO ($\bar{\nu}_\mu$)
TAU (τ^-)	1,800 TO 1,900 MeV	LESS THAN 5×10^{-12} SECOND	ANTITAU (τ^+)
TAU NEUTRINO (ν_τ)	LESS THAN 600 MeV (MAY BE 0)	NOT KNOWN (MAY BE STABLE)	TAU ANTINEUTRINO ($\bar{\nu}_\tau$)

Figure 6.2 LEPTON FAMILY has until recently been thought to consist of just four weakly interacting particles (and their corresponding antiparticles): the electron, the muon and two kinds of neutrino. The mass of the newly discovered charged heavy lepton, the tau, and its antiparticle, the antitau, is approximately 4,000 times the mass of the electron and 20 times the mass of the muon. The significance of this mass ratio is not known. Muons decay abruptly into electrons through the weak interaction. Taus decay even more quickly through the weak interaction into electrons, muons or other particles.

so remote that such a collision typically happens only once every few seconds.

In order to study all the particles that might be produced in a single electron-positron annihilation two groups of physicists from SLAC joined forces several years ago with two groups from the Lawrence Berkeley Laboratory to build a general-purpose particle detector to surround one of the interaction regions of the SPEAR device (see Figure 6.3). The detector has a cylindrical central section containing four concentric spark chambers immersed in a strong magnetic field. Surrounding the spark chambers is a system of scintillation counters for detecting charged particles. Whenever two or more charged particles are detected by the counters, the inner spark chambers are actuated and the paths of the charged particles are recorded with the aid of a computer on magnetic tape. The computer can then reconstruct the paths of the particles, yielding a diagrammatic "picture" of the event.

Outward from the scintillation counters is the aluminum magnet coil, followed by another cylindrical system of counters for detecting electromagnetic showers. The paths of particles are bent slightly by the magnetic field; the amount of bending indicates the particles momentum. These counters can distinguish electrons or positrons, which generate large showers, from hadrons or muons, which produce smaller showers. Outside all this hardware is an octagonal set of iron plates at least eight inches thick and then more spark chambers. In general hadrons cannot penetrate the iron plates because they interact with the iron nuclei through the strong force. Electrons and positrons are also unable to penetrate the iron because they have already lost most of their energy through the production of large electromagnetic showers. Muons, however, do penetrate the iron and are detected in the outer spark chambers. Hence the detector can not only measure the direction and the momentum of the newly created particles but also separately identify hadrons, electrons and muons. It was its ability to distinguish these different kinds of particles that enabled us to find the tau-antitau pair.

The SPEAR ring can be adjusted to store beams of electrons and positrons with an energy of up to four billion electron volts (GeV) in each beam. Since the collisions that occur are between matter and antimatter, a possible outcome of such a collision is what is called an annihilation reaction (see Figure 6.4). The reaction actually takes place in two steps. First the colliding electron and positron disappear,

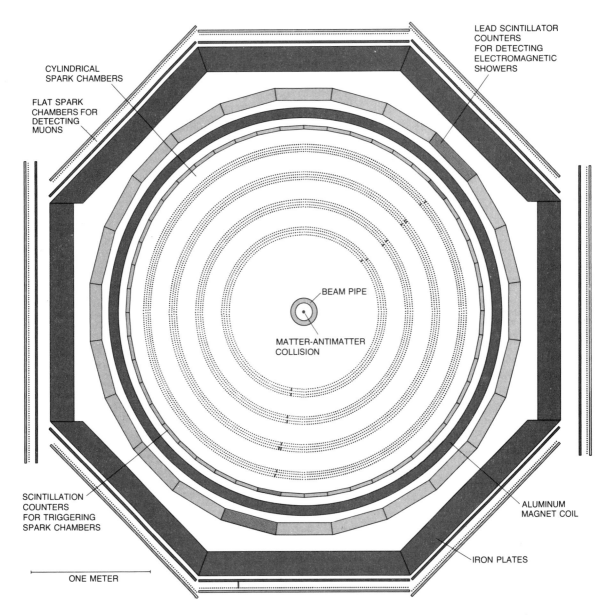

LEAD SCINTILLATOR COUNTERS FOR DETECTING ELECTROMAGNETIC SHOWERS

CYLINDRICAL SPARK CHAMBERS

FLAT SPARK CHAMBERS FOR DETECTING MUONS

BEAM PIPE

MATTER-ANTIMATTER COLLISION

SCINTILLATION COUNTERS FOR TRIGGERING SPARK CHAMBERS

ALUMINUM MAGNET COIL

IRON PLATES

ONE METER

Figure 6.3 SPEAR PARTICLE DETECTOR in cross section, perpendicular to the beam, with the collision point at center (*colored dot*). The detector has four concentric spark chambers, each with four layers of closely spaced wires (*black dots*). Spark discharges (*colored dashes*) mark the ionization trails of charged particles passing through the inert gas in the chambers. Whenever the inner system of scintillation counters detects the passage of two or more charged particles, the chambers are actuated and the paths of the particles are recorded. The outer lead scintillator counters register the distinctive electromagnetic showers generated by different types of particles. The "electron-muon" event shown in color here is interpreted in Figure 6.5.

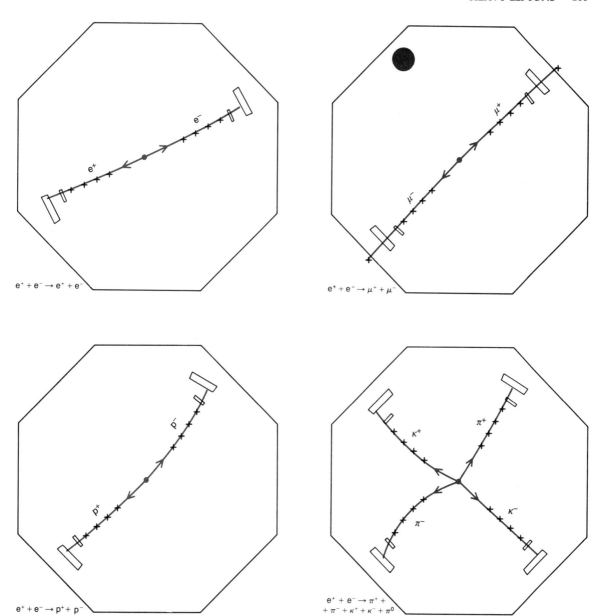

Figure 6.4 FOUR POSSIBLE OUTCOMES of the mutual annihilation of an electron and a positron. Electric discharges caused by the passage of charged particles through the spark chambers are represented by black crosses. Only those counters (*rectangles*) that have detected particles are shown. Octagonal outline indicates the outer surface of the iron plates. Colored lines connecting the detection signals are bent into circular arcs by the strong magnetic field inside the coil. In a typical electron-positron event (*top left*) the original electron-positron pair disappear and, out of the resulting energy, a new pair emerges back to back, each with half the total energy of the colliding particles. In a muon-muon event (*top right*) the muons pass freely through the iron and are detected in the outermost spark chambers. The original electron and positron annihilate each other (*bottom left*) to form a pair of hadrons, a proton and antiproton. The tracks of four hadrons, two charged pions and two charged kaons (*bottom right*), are accompanied by a neutral pion, which is uncharged and cannot be detected directly.

and in so doing they create the short-lived state of pure electromagnetic energy called a virtual photon. Then after an immeasurably brief time (10^{-25} second) the virtual photon rematerializes into any one of a very large number of possible combinations of new particles. Some of the possible outcomes are the re-creation of an electron-positron pair, the creation of a muon-antimuon pair, the creation of a hadron-antihadron pair (for instance a proton and an antiproton) or the creation of a large number of hadrons. What we proposed to do was to use the same method to produce a heavy lepton-antilepton pair.

If such pairs of new heavy leptons were actually being created in the electron-positron collisions at SPEAR, how would we be able to recognize them? Our earlier estimate of the tau's lifetime— less than a hundred-billionth of a second—meant that a newly created tau particle would travel less than a centimeter from its point of creation to the point where it decays, even if its velocity were roughly equal to that of light (30 billion centimeters per second). This path is too short to be directly detected by our experimental apparatus, and so we had to look for a way to recognize tau particles indirectly by identifying some distinctive pattern that appears when they decay.

Our decision to attribute a unique lepton number, or "taulikeness," to the tau and its neutrinos meant that we did not expect the tau to decay electromagnetically into either an electron or a muon with the accompanying emission of a photon. Instead we assumed that the tau, like the muon, would decay through the weak interaction and that several different decay modes would be possible for the tau because of its large mass. We selected two of these possible weak-decay modes for special attention.

In the first decay mode the tau decays into a tau neutrino, a muon and a muon antineutrino, and in the inverse process the antitau decays into a tau antineutrino, an antimuon and a muon neutrino. In the second decay mode the tau decays into a tau neutrino, an electron and an electron antineutrino, and the antitau decays into a tau antineutrino, a positron and an electron neutrino.

These two decay modes were picked because they were expected to occur frequently. More important, the seemingly complex decay processes described above are really very simple when they are viewed in the context of the actual experiment. The reason is that all the neutrinos and antineutrinos produced in the decay processes are such elusive particles that the experimental apparatus simply does not "see" them at all. Thus what the apparatus actually records in each case is merely the tracks of two charged particles: either an electron and an antimuon, or a positron and a muon.

This final outcome has a distinctive experimental "signature" for two reasons. First, the appearance of an electron and a muon (or their antiparticles) in the final state seems to violate the principle of the conservation of lepton number; the rule is not really violated, however, because the undetected neutrinos make up the balance. Second, there is a lot of missing energy, but that of course is also accounted for by the unseen neutrinos.

We first began to find evidence of such electron-muon events in 1974 (see Figure 6.5). In a sample of 10,000 events of all types we identified 24 as electron-muon events. Although the number of such events was low, we were encouraged, because if the events were real, it meant that the mass of the tau was within the energy range of the SPEAR system.

At that early stage, however, we had to maintain a skeptical attitude, because with just 24 events there were various ways we could be wrong. First of all, our detector was far from perfect in identifying electrons and muons. Indeed, about 20 percent of the time it misidentified a hadron as either an electron or a muon. A careful study of the problem indicated, however, that only five or six of the 24 events could possibly be attributed to a misidentified hadron among the decay products.

A second reason for skepticism was the fact that although we might have found a new particle, it might actually be a new hadron rather than a new lepton. As it happened there was at that time a prime candidate for such a new hadron: the charmed hadron called the D meson, which had not yet been discovered. It was conceivable that the electron-muon events might be coming from the production and decay of D mesons, accompanied by neutrinos and neutral K mesons. The neutral K meson is peculiar in that half of the time it decays so slowly that it would have escaped from our detector before the decay took place. In that case, if both of the neutral K mesons happened to escape, one would see only the electron and the muon in the detector and one might then be misled into thinking that the decay of a tau-antitau pair had actually been observed. As we accumulated more electron-muon events, however, particularly at higher energies, our colleague Gary J. Feldman was able to

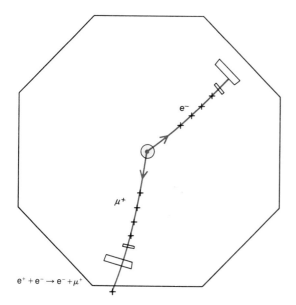

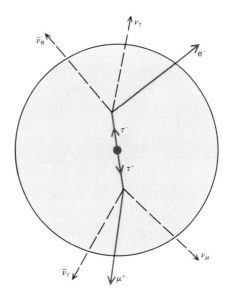

Figure 6.5 TYPICAL ELECTRON-MUON EVENT observed recently in the charged-particle detector at SPEAR is shown in the diagram at the left. The telltale "signature" of such an event is the detection of one particle that traverses the iron plates (the muon) and another that does not (the electron). The event is interpreted in the greatly enlarged view at the right in terms of the hypothesis that the electron and the muon come from the weak decay of a tau-antitau pair. The creation and decay of the heavy leptons happen within a few millimeters of the center of the detector and hence cannot be seen directly. Each tau particle decays into one charged particle and a pair of neutrinos; only the charged particles, in this case a negatively charged electron and a positively charged muon (or antimuon), are detected.

demonstrate that most of the electron-muon events could not have come from the production of *D*-meson pairs, because there were too few cases in which we observed both the electron-muon signal and the expected decay products of the neutral *K* mesons.

In the two years following our initial discovery we continued to collect more electron-muon events (we now have about 200) and to look for other ways of testing our data. One important test was to see whether the energy distribution of the electron and the muon was what one would expect from the weak decay of a heavy lepton into three particles. We found that the data did fit the three-body hypothesis well but that it did not fit the alternative two-body hypothesis (see Figure 6.6).

In another test we considered other possible decay modes of the tau-antitau pair. If, for example, one of the tau particles were to decay into a meson (mesons are a subclass of the hadrons) and the other were to decay into a muon, then one would expect to see distinctive muon-meson events. Such events would be distinctive because, like the electron-muon events, they would seem to violate the principle of the conservation of lepton number. A group of physicists from the University of Maryland, the University of Pavia and Princeton University found a few such muon-meson events in an experiment at SPEAR, and later our Stanford-Berkeley group was able to collect about 100 similar events.

Certain other links remained to be closed in the chain of evidence. The German Electron Synchrotron Laboratory in Hamburg has an electron-positron storage ring called DORIS, which has a particle-producing capability similar to SPEAR's. If tau particles were being created in the Stanford device, they should also be created in the Hamburg one. For a year or so there were no reports of heavy leptons being found at DORIS, but then a group at the Hamburg facility began to observe electron-muon events and later muon-meson events; in both cases they had the characteristics expected for the tau-antitau pair. Additional confirmation was provided

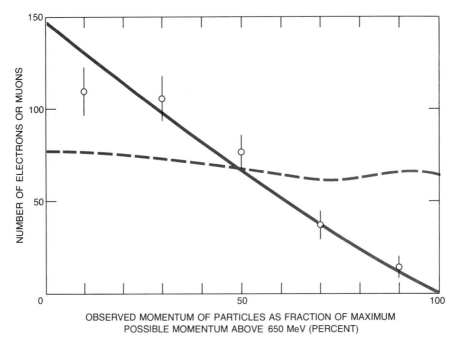

Figure 6.6 EVIDENCE that the newly discovered elementary particle is a lepton and not some novel kind of hadron is presented in this graph of the energy distribution of the particles observed emerging from the electron-muon events. Assuming that the tau is a heavy lepton that decays weakly into three particles (either an electron and two neutrinos or a muon and two neutrinos), the data would be expected to follow the solid colored curve. If the tau were another kind of particle that decays into an electron or a muon, in either case accompanied by a single undetected neutral particle, then the data would follow the broken colored curve.

when another group at DORIS discovered a number of electron-meson events, which are analogous to the muon-meson events.

At an international meeting on the physics of leptons and photons held last August in Hamburg five experimental groups working at SPEAR and two groups working at DORIS reported the results of their independent searches for heavy-lepton production in electron-positron annihilation. All the groups agreed on the following points: (1) that they had evidence for a new particle, (2) that the new particle was not a charmed hadron, (3) that the mass of the new particle was between 1,800 and 1,900 MeV (close to 20 times the muon's mass and 4,000 times the electron's) and (4) that all the measured properties of the new particle were consistent with the properties one would expect for a heavy charged lepton.

One way of showing that the new particle is a lepton and not a hadron is to measure its rate of production as a function of the total energy of the electron-positron collision. From studies of the pro-

duction of pairs of hadrons it is known that when the total energy of the system is raised above a certain threshold energy, the production rate first increases for a time and then begins to decrease rapidly. The rapid decrease ensues because when there is too much energy available, the hadrons, which are composite particles, cannot hold together. Instead of a pair of hadrons many hadrons are produced. In contrast, since leptons are pointlike particles that do not break up, one would expect the production rate of leptons to rise rapidly from the threshold energy, reach a maximum and then decrease rather slowly as the energy is increased. Using the electron-muon events as a measure of the production rate of tau-antitau pairs, we have determined that the production rate of the new particles changes with increasing energy in the manner predicted for lepton pairs and not in the manner predicted for hadron pairs (see Figure 6.7).

Although all the experiments to date agree that there is a new heavy lepton, there is still much we do not know about it. For example, the existing data

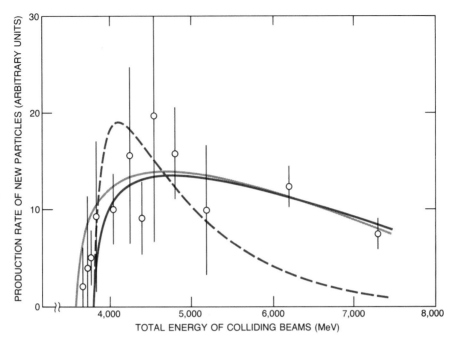

Figure 6.7 FURTHER EVIDENCE that the tau is a lepton and not a hadron is presented in this graph, in which the rate of production of the new particle is plotted as a function of the total energy released in the electron-positron collision. If the new particles were hadrons, then as the total energy of the system is raised above a certain threshold energy their production rate would increase for a time and then decrease rapidly (*broken colored curve*). If the tau particles were leptons, their production rate would rise rapidly from threshold energy, reach a maximum and then decrease slowly as the energy is increased (*solid colored curves*). The dark colored curve is based on the assumption that the mass of the new lepton is 1,900 MeV; the light colored curve 1,800 MeV.

are consistent with the idea that the tau is not affected by the strong force, but this insensitivity has not been tested as thoroughly for the tau as it has been for the electron and the muon. In addition many of the elementary particles, including the leptons, are in a state of perpetual rotation about their own axes, like spinning tops, but it has not yet been established that the tau has the same spin characteristics as the electron and the muon. We also do not know yet whether the tau is the sequential heavy lepton we initially set out to find. In other words, does the tau have its own unique tau lepton number, or taulikeness, or is it an entirely new kind of lepton with even more peculiar properties?

Experiments at other laboratories based on the interactions of muon neutrinos do not seem to have created any tau particles; hence it is unlikely that the tau is muonlike. Nevertheless, there remains the possibility that it is electronlike. (If it is, however, there must be a rather special and complicated mechanism at work to suppress the expected elec-

tromagnetic decay of the tau into an electron and a photon.) Furthermore, not all the possible decay modes of the tau have been observed, and it is only through knowing the full set of decay modes that one can know just what kind of lepton the tau is.

Another area that will require further experimental and theoretical study is the question of why the mass of the tau (between 1,800 and 1,900 MeV) is so close to the estimated mass of the charmed D meson (1,865 MeV), particularly when there is strong evidence that the tau is itself not a charmed hadron. As it happens, there is another such puzzling correspondence in particle physics: the mass of the muon (106 MeV) is quite close to the mass of the pi meson, or pion (140 MeV). Are these merely chance coincidences or is there some unknown relation between the masses of the leptons and the hadrons?

There are other questions. For example, what determines the masses of the leptons? It is hard to make sense of the observed mass sequence of the electron, the muon and the tau: .51, 106 and 1,800 to 1,900 MeV. The numbers increase too quickly to be an arithmetic series and too slowly to be a geometric series. Of course, given only three points, there are many empirical formulas that could be found to fit the sequence, but none of them would be based on a fundamental understanding of the leptons, since no one knows what accounts for the individual masses of the leptons.

And what is one to make of the properties electronlike, muonlike and taulike? Perhaps there is nothing further to understand about these properties. Just as physicists have come to accept electric charge as a fundamental property of particles for which they have no deep understanding, so too they may simply have to accept the unique lepton properties electronlike, muonlike and taulike. It is still not known why the total electric charge is conserved in all particle interactions; perhaps it is not possible to know why the total lepton numbers are conserved in all lepton reactions.

Answers may nonetheless be in the offing. Two new high-energy electron-positron colliding-beam accelerators are now under construction: the PEP machine at SLAC and the PETRA machine in Hamburg. Both devices will achieve an energy of 18,000 MeV or more per beam, which means that the search for additional charged leptons can be extended up to masses perhaps five times greater than the present upper limit of about 3,500 MeV. Most of the experiments being planned for PEP and PETRA include in their design ways to search for new heavy leptons, usually by looking for electron-muon events. Whether or not new leptons are found it is certain that the search will be more difficult, if only because the electron-muon events from the decay of the tau particles will be an annoying background. Experimenters will have to distinguish new and interesting electron-muon events from old and uninteresting ones resulting from the tau-anti-tau decay. On the other hand, the electron-muon events from the taus will be helpful in checking out the new devices.

Finally, there is a deeper question for which no answer exists at present. What is the relation between the two multimember classes of particles that now seem truly elementary, the leptons and the quarks? Before the discovery of the tau particles there were only four known types of lepton and four known types of quark (counting each particle and its antiparticle as one type). There was a nice

symmetry here, and certain theories relating the leptons and the quarks made use of it. With the discovery of the new heavy lepton, however, the symmetry is destroyed; there are now more known leptons than there are known quarks, two more if the tau is sequential with its own tau neutrino.

Meanwhile, however, a group of physicists at the Fermi National Accelerator Laboratory (Fermilab) has recently reported evidence of the possible existence of a new, fifth kind of quark. Thus the number of quarks may also be increasing. Some theories preserve the symmetry between the leptons and the quarks, others forgo the need for such a symmetry, but almost all provide for the possibility of additional quarks and leptons. This apparent proliferation in the types of leptons and quarks, although still a matter of speculation, is somewhat alarming. In many ways it would be preferable if the truly elementary particles could remain as few in number as they were in the pre-tau days, or better yet in the pre-tau and pre-charmed-quark days. One cannot, however, dictate to nature what the fundamental constituents of matter should be. One can only hope to be up to the task of finding the constituents and understanding them.

POSTSCRIPT

In our article we described the discovery of the tau lepton and suggested four problems for further study: What are the detailed properties of the tau? Are there charged leptons heavier than the tau? What determines the masses of the leptons? Are leptons and quarks related and how do they fit into the overall pattern of elementary particles and fundamental forces?

The answers are a mixed lot and we'll take them in turn. Many measurements have been made on the properties of the tau. It is a sequential lepton, that is, it behaves like a heavy electron or a heavy muon. Its large mass (1784 ± 3 MeV) compared to the other leptons allows it to decay in some dozen different ways. Besides decaying to an electron and neutrinos or to a muon and neutrinos, as we discussed, it also decays into pi or K mesons plus neutrinos. To the best of our knowledge one of those neutrinos is always a tau-neutrino.

Almost all the ways by which the tau lepton can decay have been seen, and their measured properties agree with calculations using the modern theory of the weak interaction, the Weinberg-Salam theory as described in Chapter 1 by Chris Quigg. The decay lifetime of the tau, several tenths of a picosecond (about a million-millionth of a second), is explained by these calculations. Indeed, except for 5 or 10 percent of the decays that we do not yet understand, the tau is a textbook example of how a heavy charged lepton should behave. Thus, very satisfying progress has been made in answering the first question.

As to the second question on the possibility of additional charged leptons, the answer so far is negative and unsatisfactory. It is negative because no additional charged leptons have been found in searches with electron-positron colliding beams accelerators for masses up to 22 GeV and in preliminary studies with proton-antiproton collisions up to 40 GeV, 22 times the tau mass. Searches have also been made for additional neutral leptons, that is, in addition to the electron, muon and tau neutrinos, and although less comprehensive, nothing has been found.

That no additional leptons have been found is unsatisfactory because no quantitative answer has been found for the third question of what sets the masses of the leptons? There are many ideas expressed in Chapter 1 by Quigg, Chapter 12 by Haim Harari and by Howard Georgi [see "A Unified Theory of Elementary Particles and Forces"; SCIENTIFIC AMERICAN, April, 1981] but no usable and believable equation for the lepton masses has resulted. Therefore, if no other lepton is found, we cannot know if that null result is significant.

The fourth question on the place of the leptons in the general scheme of particles and forces is also unanswered. Again, there are many ideas but none have been proven.

The Search for Intermediate Vector Bosons

In theory these massive elementary particles are required to serve as the carriers of the weak nuclear force. They should be detected soon in the aftermath of collisions between protons and antiprotons.

. . .

David B. Cline, Carlo Rubbia and Simon van der Meer
March, 1982

One of the major achievements of modern physics has been the development over the past 15 years or so of a new class of unified theories to describe the forces acting between elementary particles. Until these theories were introduced the four observable forces of nature seemed to be quite independent of one another. The electromagnetic force governs the interactions of electrically charged particles; the weak nuclear force is responsible for such processes as the beta decay of a radioactive atomic nucleus; the strong nuclear force holds the nucleus together, and gravity holds the universe together. The most successful of the new theories establishes a link between electromagnetism and the weak force, suggesting that they are merely different manifestations of a single underlying force.

The unified electroweak theory is about to be put to a decisive experimental test. A crucial prediction of the theory is the existence of three massive particles called intermediate vector bosons (also known as weakons). The world's first particle accelerator with enough energy to create such particles has recently been completed at the European Organization for Nuclear Research (CERN) in Geneva. The accelerator was originally built to drive high-energy

protons against a fixed target, but it has now been converted to a new mode of operation, in which protons and antiprotons collide head on. The colliding-beam machine has completed its preliminary runs and will resume operation with an elaborate new particle-detection system in place (see Figure 7.1). If the intermediate vector bosons exist and if they have the properties attributed to them by the electroweak theory, they should be detected soon. They are currently the most prized trophies in all

Figure 7.1 NEW PARTICLE DETECTOR at CERN was designed and built by a team of more than 100 physicists from 11 institutions in Europe and the U.S. The first observation of an intermediate vector boson is expected to be made with the aid of this device. The large, multipurpose detector, named the UA1 (for Underground Area One), is seen here in its "garage" adjacent to the largest particle accelerator at the CERN site: the Super Proton Synchrotron (SPS), which was recently converted into a proton-antiproton colliding-beam machine. When the detector is ready for operation, it is rolled to the left on rails into the path of the colliding beams. Electronic equipment covers the outside of the apparatus, obscuring the central detection chambers. (Photo by European Organization for Nuclear Research.)

physics, and their discovery would culminate a search that began more than 40 years ago.

According to the prevailing view of the interactions of elementary particles, a force is transmitted between two particles by the exchange of a third, intermediary particle. Such a description is the essence of a quantum field theory. The idea of a field extending throughout space is needed to explain how particles can act on one another at a distance; it is a quantum field because it is embodied in discrete units, namely the intermediary particles. In electromagnetic and weak interactions the exchanged particle is a member of the family called the vector bosons. This term refers to a classification of particles according to one of their most basic properties: spin angular momentum. A boson (named for the Indian physicist S. N. Bose) is a particle whose spin, when measured in fundamental units, is an integer, such as 0, 1 or 2. "Vector" designates a boson whose spin value is equal to 1.

In the case of electromagnetism the exchanged vector boson is the photon, the massless and chargeless "wave packet" of electromagnetic energy that functions as the quantum of the electromagnetic field. Photons are easy to observe experimentally (in the form of light, for example), and from the study of their properties physicists have constructed the remarkably precise and comprehensive theory called quantum electrodynamics, or QED, which is the quantum field theory of the electromagnetic force.

The corresponding force carrier in weak interactions is the intermediate vector boson (intermediate simply because of its mediating role between particles). The existence of such a particle was first suggested in 1935 by the Japanese physicist Hideki Yukawa, who at the time was seeking a unified explanation of the two newly discovered nuclear forces: the strong and the weak. Yukawa noted that the range of a force should be inversely proportional to the mass of the particle that transmits it. For example, the range of the electromagnetic force is infinite, in accord with the masslessness of the photon. On the other hand, the two nuclear forces have only a limited range, and so Yukawa suggested they should be carried by particles with mass.

Specifically, Yukawa postulated the existence of a moderately heavy particle, later named the meson, the exchange of which gives rise to the strong attractive force between the proton and the neutron. The first particle of this type to be correctly identified, the pi meson (or pion), was discovered in 1947 in the shower of secondary particles generated by the collision of a cosmic-ray particle with an atom in the atmosphere; large numbers of mesons can now be made at will with particle accelerators.

The particles of the nucleus, including the mesons, are now thought to be composed of the more fundamental constituents known as quarks. The quarks are bound together by the strong force, but in this context the force has a form quite different from the one observed between protons and neutrons. The force between quarks is thought to be transmitted by the family of eight massless vector bosons called gluons. A property with the arbitrary name "color" is assigned to the quarks and the gluons and plays the same role in strong interactions as electric charge does in electromagnetic interactions. In recognition of this analogy the quantum field theory of the strong force is called quantum chromodynamics, or QCD.

T he weak nuclear force has a still shorter range than the strong force that acts between protons and neutrons. The intermediate vector bosons of the weak force can therefore be expected to have a mass larger than that of the pi meson. Early attempts to detect the intermediary particles associated with the weak force were unsuccessful, presumably because the bosons' larger mass put them out of reach of the existing particle accelerators. Until the advent of the unified electroweak theory in the late 1960's, however, there was no good estimate of the mass of the weak-force particles.

The electroweak theory was developed independently by Steven Weinberg of Harvard University and Abdus Salam of the International Center for Theoretical Physics in Trieste, with major contributions by Sheldon Lee Glashow of Harvard and others. The theory, which can now be considered the "standard" account of electromagnetic and weak interactions, for the first time made specific and testable predictions about the properties of the intermediate vector bosons, including their mass. Furthermore, the theory required that there be three such particles, with electric charges of +1 (the W^+), −1 (the W^-) and zero (the Z^0). The best present estimate of the mass of the intermediate vector bosons, expressed in terms of their equivalent energy, is 79.5 GeV for the W^+ and W^- and 90 GeV for the Z^0. (The abbreviation GeV stands for gigaelectron volts, or billions of electron volts; for comparison the mass of the proton is equivalent to a little less than one GeV.)

The idea at the heart of the standard theory is that electromagnetism and the weak force both stem from a single and more fundamental property of nature. At exceedingly high energy (high enough for W and Z particles to be made as readily as photons) events mediated by the two forces would be indistinguishable. This theoretical unification is accomplished by assigning the photon and the intermediate vector bosons to the same family of four particles. At energies accessible today there is no question that electromagnetic events are quite different from weak ones; moreover, the photon and the W and Z particles seem to be unlikely siblings, since the first is massless and the other three are among the heaviest particles supposed to exist. The discrepancy is explained in the standard theory by the notion of a broken symmetry, which distinguishes between the forces as the energy is lowered in much the same way as a substance separates into different phases as the temperature is lowered.

One approach to understanding the unified electroweak theory begins with an imaginary primordial state in which the photon and the intermediate vector bosons were all equally massless. It was the breaking of a symmetry of nature that endowed the W^+, the W^- and the Z^0 with large masses while leaving the photon massless. A mechanism for discriminating in this way among the carriers of the forces was first discussed in 1964 by Peter Higgs of the University of Edinburgh. Curiously, the Higgs mechanism is able to supply masses for the W and Z particles only by postulating still another massive particle, which has come to be called the Higgs boson. It is being sought along with the intermediate vector bosons.

The electroweak theory received important experimental support in 1973 from a discovery made at CERN and at the Fermi National Accelerator Laboratory (Fermilab) near Chicago. Up to then all known weak interactions of matter entailed an exchange of electric charge. For example, a proton might give up its charge of $+1$ to a neutrino (a massless particle with no charge). As a result the proton becomes a neutron and the neutrino is converted into a positron, or antielectron. All such events can be accounted for by the exchange of the charged intermediate vector bosons W^+ and W^-. The 1973 experiments revealed weak interactions in which the particles maintain the same charges they had before the event, as they do in electromagnetic interactions. A weak interaction of this type can be explained only by the exchange of a neutral intermediate vector boson (the Z^0 particle), or, in an equivalent description, by the operation of a neutral weak current [see "The Detection of Neutral Weak Currents," by David B. Cline, Alfred K. Mann and Carlo Rubbia; SCIENTIFIC AMERICAN, December, 1974]. In 1979 Weinberg, Salam and Glashow were awarded the Nobel prize in physics "for their contributions to the theory of the unified weak and electromagnetic interaction between elementary particles, including . . . the prediction of the weak neutral current.

Once the existence of neutral weak currents had been fully confirmed it was only natural to try to find a way to detect the Z^0 as well as the W^+ and W^-. The task of creating particles with such a large mass, however, remained daunting. The largest particle accelerators at the time were machines in which a single beam of protons is raised to high energy and then directed onto a fixed target. In the ensuing collision of a beam particle with a target particle most of the energy released goes into moving the two-particle system rather than demolishing it; only a small fraction of the beam energy is made available for the creation of new particles. The only chance of observing an intermediate vector boson, it seemed clear, would be in a colliding-beam machine, where the accelerated particles meet head on, transforming essentially all their energy into new particles.

Storage rings for electrons and positrons had already been in operation for several years. The great advantage of employing electrons and positrons is that a single ring of magnets and radio-frequency cavities can simultaneously accelerate a particle and its antiparticle in opposite directions, so that counterrotating beams are formed in a single doughnut-shaped vacuum chamber. On the other hand, because electrons and positrons are very light they rapidly dissipate their energy when they are made to follow the curved path of the storage ring. It did not seem feasible then to build an electron-positron machine large enough to reach the energy of the intermediate vector bosons. The plan instead was to build storage rings in which protons would collide head on with other protons; two interlaced rings were needed to arrange such collisions. The first of these proton-proton machines were not scheduled to begin operating until the mid-1980's or later.

Then in 1976 two of us (Cline and Rubbia), together with Peter M. McIntyre, came up with an alternative idea. Instead of building an entirely new colliding-beam machine, we proposed, it would be feasible (and much cheaper) to convert an existing fixed-target proton accelerator into a colliding-beam machine by arranging to generate a counterrotating beam of antiprotons in the same annular space occupied by the original proton beam. Our suggestion was well received, and after a thorough review of the problems likely to be encountered in such a project it was decided to build proton-antiproton machines at two of the world's largest proton accelerators: the Super Proton Synchrotron (SPS) at CERN, which began operating at a peak energy of 400 GeV in 1976, and a more advanced version of a comparable machine at Fermilab that was then still in the planning stage.

The CERN conversion was in many ways easier to accomplish, and it was completed by last summer, under the direction of Roy Billinge and one of us (van der Meer); the first proton-antiproton collisions at the desired peak energy of 270 GeV per beam were observed in July. By late December, when the machine was shut down for the Christmas holiday, more than 250,000 such collisions had been recorded; because of the comparatively low rate at which intermediate vector bosons are expected to be produced in proton-antiproton collisions, however, it was not surprising that none were detected in these early runs. This situation is expected to change dramatically in the next round of experiments, in which the particle intensity of the beams, and hence the collision rate, will be increased by an order of magnitude or more.

The big proton-antiproton colliding-beam machine at Fermilab is still under construction and is scheduled to begin operating in 1985. Because it was originally designed to accelerate a single beam of protons to an energy of 1 TeV, or a trillion electron volts, it was named the Tevatron. In its reincarnation as a colliding-beam machine it is expected to be able to produce collisions with a total energy of 2 TeV (2,000 GeV), as opposed to 540 GeV for the CERN machine. When the Fermilab machine is completed, it will have the further distinction of being the first large accelerator to employ a ring of superconducting magnets.

How does one go about generating opposed beams of matter and antimatter in a storage ring? The hardest part in the two present instances is to accumulate a dense enough "bunch" of antiprotons to ensure a large number of collisions with the counterrotating protons. Unlike protons, antiprotons are not readily available from any natural source; they must themselves be created in high-

energy collisions. A beam of high-energy protons is directed at a metal target, and antiprotons created in collisions with the target atoms are steered magnetically into a specially designed storage ring. The process is extremely inefficient; on the average one comparatively low-energy antiproton is produced for every million or so high-energy protons striking the target. To put this production rate in perspective, it has been calculated that in order to obtain a useful number of proton-antiproton collisions in the colliding-beam machine at CERN one must collect bunches of antiprotons (and protons) each made up of at least 100 billion particles. Successive bunches of antiprotons are collected and "stacked" every 2.4 seconds; at this rate it takes about 24 hours to accumulate a few hundred billion antiprotons for the colliding-beam experiments at CERN (see Figure 7.2).

Creating enough antiparticles is not the only problem. As the antiprotons emerge from the target they have a range of velocities and directions. Viewed in their own frame of reference the antiprotons form a gas, and their random motions are indicative of a temperature. If the temperature is too high, some of the particles will strike the walls of the accelerator and the beam will be dissipated. Therefore some method is needed to "cool" the antiproton beam (that is, to reduce its random motions) in order to keep it as concentrated as possible before it enters the accelerator ring.

One such beam-cooling technique, called electron cooling, was first proposed more than a decade ago by Gersh I. Budker of the Institute of Nuclear Physics at Novosibirsk in the U.S.S.R. Basically it operates by mixing a "cool" beam of electrons (one in which all the particles have the same velocity and direction) with the "hot" antiproton beam for a short distance. In the process some of the random thermal energy of the antiprotons is transferred to the electrons. Mixing the antiproton beam repeatedly with fresh electron beams can cool the antiprotons significantly, provided their energy is not too high to start with. As it happens, the CERN scheme calls for antiproton beam whose energy is initially too high to be cooled effectively by this method, and it is no longer being studied for this purpose. Electron cooling is still under consideration, however, for a role in the Fermilab project.

Another beam-cooling method, better suited to the requirements of the CERN proton-antiproton machine, was invented in 1968 by one of us (van der Meer). This method, called stochastic cool-

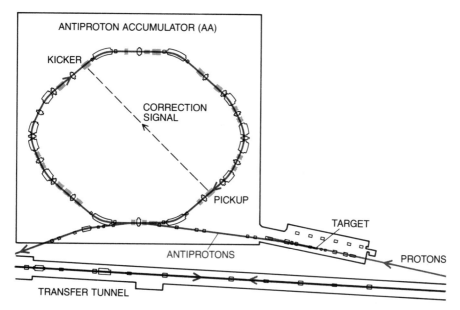

Figure 7.2 ANTIPROTON ACCUMULATOR in the CERN colliding-beam experiments "stacks" the successively injected bunches of antiprotons, and "cools" them by a statistical process known as stochastic cooling. For cooling the ring incorporates a number of linked "pickup" and "kicker" devices (*gray shapes*) as well as the beam-bending and beam-focusing magnets (*white shapes*) found in any storage ring. In stochastic cooling a pickup at one section of the storage ring senses the average deviation of the particles from the ideal orbit; a correction signal is then sent across the ring to a kicker on the other side, arriving just in time to nudge the particles back toward the ideal orbit.

ing (because it relies on a statistical process), utilizes a "pickup," or sensing device, in one section of a storage ring to measure the average deviation of the particles from the ideal orbit. The measurement is converted into a correction signal, which is relayed across the ring to a "kicker" device on the other side. The kicker applies an electric field to its section of the ring in time to nudge the center of mass of the passing particles back toward the ideal orbit. Although the particles move with nearly the speed of light, the correction signal can arrive in time because it takes the shorter path across a chord of the cooling ring.

Both beam-cooling techniques have been tested successfully in the past few years at Novosibirsk, CERN and Fermilab. As a result there is every reason to believe full-scale antiproton collector rings such as the ones at CERN and Fermilab will work as planned. Beam cooling is becoming a routine part of accelerator technology.

For the experiments at CERN the particles are directed through a complex sequence of interconnected beam-manipulating devices (see Figure 4.8). First a beam of protons is accelerated to an energy of 26 GeV in the Proton Synchrotron (PS), the original accelerator ring at CERN, completed in 1959. The proton beam is then directed at a copper target, producing a spray of particles, including a small number of antiprotons with an energy of 3.5 GeV. The antiprotons are collected and transferred to a wide-aperture storage ring called the Antiproton Accumulator (AA), where they are first precooled by the stochastic method and then moved to a slightly smaller orbit, where they are stacked with the previously injected bunches and subjected to further cooling (see Figure 7.3). After a few hundred billion antiprotons have been collected they are sent back to the PS ring, where they are accelerated to 26 GeV before being injected into the SPS. Meanwhile protons at 26 GeV from the PS ring are injected into the SPS ring in the opposite direction. The counter-rotating beams are finally accelerated to 270 GeV each in the SPS ring. The beams collide at two interaction sites, where the large particle detectors are placed. The interactions are so rare that the beam lifetime of several hours is not affected by them.

At present the Fermilab plan calls for a more

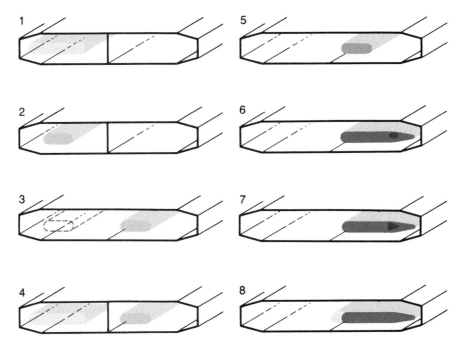

Figure 7.3 STACKING AND COOLING of antiprotons (shown in cross-sectional diagrams) begins when about 20 million antiprotons are injected into the accelerator ring and made to circulate on the outside (*left*) of the vacuum chamber (*1*). The injected particles are precooled for 2 seconds, reducing their random motion (*2*). A mechanical shutter dividing the chamber is lowered and the antiprotons are magnetically shifted into the main body (*3*). The shutter is raised and a second bunch of antiprotons is injected (*4*) and subjected to the same procedure (*5*). About an hour later a dense core begins to form in the stack (*6*). After 40 hours a trillion antiprotons are orbiting in the stack (*7*). Magnetic fields are then actuated to extract the core, providing about 600 billion antiprotons for experiments. The residue of some 400 billion antiprotons remains stacked, available for the next core (*8*).

intense antiproton source than the one in operation at CERN. Protons with a higher energy will be used to make the antiprotons at Fermilab. Several alternative schemes are being investigated. One would rely on a combination of stochastic cooling and electron cooling to accumulate 100 billion antiprotons in less than an hour. The antiprotons would be injected into the Tevatron ring, accelerated to an energy of 1,000 GeV and made to collide with counterrotating 1,000-GeV protons in two experimental areas. The first area, where construction is scheduled to start soon, is designed to house a very large detector.

According to the electroweak theory, intermediate vector bosons can be created in proton-antiproton collisions by a variety of mechanisms in which a quark from the proton's structure interacts with an antiquark from the antiproton's structure. Both the proton and the antiproton are assumed to be made up of three constituent particles; in the whimsical nomenclature of the QCD theory the proton is said to harbor two "up" quarks (labeled u) and one "down" quark (d), whereas the antiproton has two "antiup" antiquarks ($\bar{u}$) and one "antidown" antiquark ($\bar{d}$). When a quark and an antiquark collide, they annihilate each other, creating a burst of energy that can rematerialize as new particles, including intermediate vector bosons. In some cases a single intermediate vector boson is expected to appear (accompanied by other kinds of particles); in other cases a pair of intermediate vector bosons is predicted.

The expected production rate of W^+, W^- and Z^0 particles in proton-antiproton collisions varies as a function of an experimental parameter called the luminosity, which is defined as the number of high-energy particles per square centimeter per second passing through the cross section of the interaction region. The designed luminosity of the CERN machine, assuming the injection of 600 billion antiprotons per bunch, is 10^{30} particles per square centimeter per second. Given the same number of antiprotons per bunch, the Fermilab machine should attain a luminosity of 4×10^{30} particles per square centimeter per second (owing to its higher energy and hence smaller beam size). It can be calculated that at such luminosities the production rate of W^+, W^- and Z^0 particles, both singly and in pairs, should be high enough for them to be detected fairly often, perhaps as often as thousands of times per day (see Figure 7.4). Just before the CERN machine was shut down in 1982 it briefly reached a luminosity of 10^{28} particles per square centimeter per second, putting it at the threshold of where one would expect to catch the first glimpse of the inter-

mediate vector boson. (The initial operating experience with the CERN collider has also shown that proton-antiproton collisions could in principle be obtained at luminosities of 10^{31} particles per square centimeter per second or higher, provided enough antiprotons are available.)

How will intermediate vector bosons produced in such collisions make their presence known? The lifetime of the particles is expected to be exceedingly short. In approximately 10^{-20} second they should decay spontaneously to form a variety of other particles, mainly quark-antiquark pairs and lepton-antilepton pairs. (Leptons are particles that respond to the weak nuclear force but not to the strong one.) Charged leptons such as electrons and muons can be detected by various means. In general the object

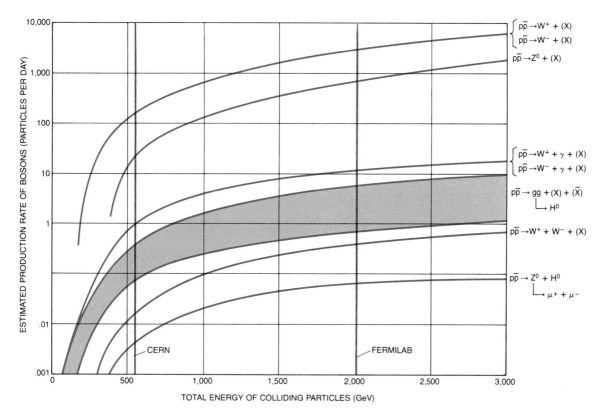

Figure 7.4 ESTIMATED PRODUCTION RATES of intermediate vector bosons and Higgs bosons by various collision processes. The curves indicate the rate calculated for each process for one day of operation of the CERN and Fermilab colliding-beam machines. The two vertical gray lines indicate the designed total energy of the proton-antiproton collisions at CERN and Fermilab. The processes that result in the production of a charged intermediate vector boson accompanied by an energetic photon are expected to serve as sensitive indicators of the magnetic properties of the W^+ and W^- particles. The process in which a Higgs boson is produced by the fusion of two gluons has a less predictable rate shown by a range of values.

is to detect the charged leptons from the decay of the intermediate vector bosons and to compare their production rate or other properties with the values expected from other causes arising in the collision process. The signal, in this case the angular distribution of leptons from the decay of W^+, W^- or Z^0 particles, is expected to stand out above the background "noise" of leptons from other sources, particularly at large angles with respect to the beam axis. In Figure 7.5 notice that for the W case (*left graph*) the background rate of leptons expected from other sources related to the collision process has no peak and is lower than the rate expected from the decay of the W^+ and W^- particles. For the Z case (*right graph*) the situation is even better. The spectrum has a sharp peak near the predicted mass of the Z^0 particle (90 GeV). The negligible background noise makes this process the one in which the Z^0 is most likely to be discovered.

One unmistakable indication of the presence of intermediate vector bosons would be the appearance of a marked asymmetry in the rate at which leptons are detected in the forward and backward directions (measured arbitrarily with respect to the direction of the antiproton beam). Leptons that arise directly from strong or electromagnetic interactions

of the beam particles should be completely symmetrical. According to the electroweak theory, however, decaying intermediate vector bosons should emit positively charged leptons predominantly in the forward direction and negatively charged ones predominantly in the backward direction. The expected lepton asymmetry, which is unique to events mediated by the weak force, results from the spins of the particles involved in the production and decay of intermediate vector bosons (see Figures 7.6 and 7.7). The observation of this effect will be taken as strong evidence that the long-sought intermediate vector bosons have finally been discovered. Their other properties can then be measured.

The ultimate test of the correctness of the electroweak theory would be the observation in the debris of proton-antiproton collisions of the Higgs boson. The discovery of this particle would demonstrate not only that electromagnetism and the weak force are unified but also that the unification is of the kind prescribed by the standard electroweak theory. A full discussion of the experimental technique required to detect the Higgs boson is beyond the scope of this article. Proton-antiproton collisions could give rise to Higgs bosons, however, and the estimated rate of production is high enough for

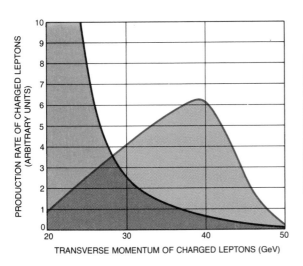

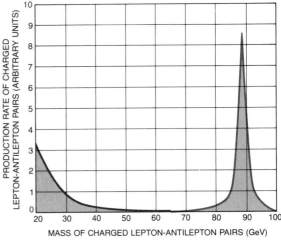

Figure 7.5 TELLTALE SIGNALS of the production of intermediate vector bosons (*color*) are expected to stand out above the background "noise" (*gray*). The graph at the left shows the calculated mass spectrum of the charged leptons that would be emitted with a large transverse momentum from proton-antiproton collisions in which charged intermediate vector bosons are created. The peak in the signal is predicted to appear at about half the estimated mass of a charged intermediate vector boson. The graph at the right shows the calculated mass spectrum for the decay of a neutral intermediate vector boson into a pair of charged leptons. The spectrum in this case has a peak near the predicted mass of the Z^0 particle (90 GeV).

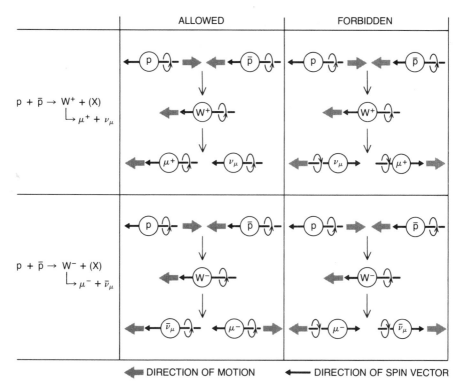

Figure 7.6 DIRECTION OF SPIN of the particles involved in the production and decay of intermediate vector bosons has an important effect on the direction of motion of the decay products. The result is an asymmetry in the rate at which charged leptons are expected to be observed in different directions with respect to the incoming beams. For example, when a charged intermediate vector boson (W^+ or W^-) decays to form a muon (μ^+ or μ^-) and a muon-type neutrino (ν_μ or $\bar\nu_\mu$), most of the positively charged muons leave the collision in the direction of the incoming antiproton beam, whereas most of the negatively charged muons leave in the direction of the incoming proton beam (here left and right respectively).

them to be discovered by means of their characteristic decay products in either the CERN or the Fermilab colliding-beam machines (see Figure 7.8).

Several large detectors have been designed to search for the decay products of intermediate vector bosons and Higgs bosons. One of these devices, named the UA1 (see Figure 7.9), is now finished and ready for the resumption of the search at CERN. The detector is the result of a collaborative effort by a team of more than 100 physicists from 11 institutions in Europe and the U.S.: the University of Aachen, the Annecy Particle Physics Laboratory, the University of Birmingham, CERN, Queen Mary College (London), the Collège de France (Paris), the University of California at Riverside, the University of Rome, the Rutherford Laboratory, the Saclay Nu-

clear Research Center and the University of Vienna. It is 10 meters long by five meters wide, and its total weight is 2,000 tons. The hall, 25 meters underground, in which it is installed is large enough for the detector to be rolled back into a "garage" when it is not in place in the path of the colliding beams.

The UA1 detector is a multipurpose device, designed to sense many kinds of particles and to collect information over a wide solid angle surrounding the point where the beams collide. It measures the energy of the particles by several means, including the curvature of their paths in a magnetic field. A large dipole magnet applies the main magnetic field horizontally throughout a volume of 85 cubic meters.

Inside the magnet, surrounding the beam tube, there are three "drift chambers," each containing an

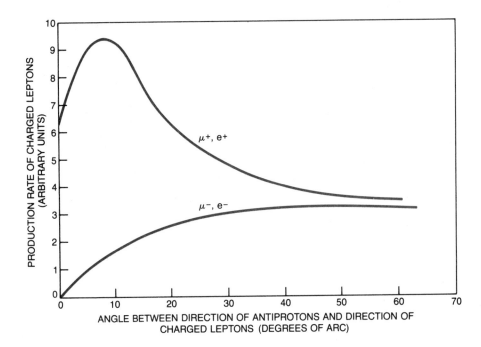

Figure 7.7 ASYMMETRY IS PREDICTED in the angular distribution of charged leptons emerging from proton-antiproton collisions in which charged intermediate vector bosons are created. Here the theortical production rate of the charged leptons is given at various angles in relation to the direction of the antiproton beam. The effect is strongest in the forward direction (that is, at small angles with respect to the antiproton direction). Electrons (e^-) and positrons (e^+) as well as muons are expected to contribute to the effect. The data points for the two curves were calculated for proton-antiproton collisions at a total energy of 2,000 GeV.

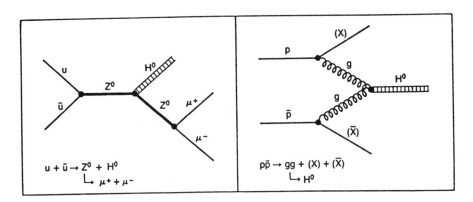

Figure 7.8 HIGGS BOSON might also make its first appearance in the colliding-beam experiments at CERN or Fermilab. The discovery of this massive, uncharged particle (designated H^0) is considered the ultimate test of the "standard" unified theory linking electromagnetic interactions and weak interactions. At the left a Higgs boson is created in association with a neutral intermediate vector boson. At the right a Higgs boson arises from the fusion of two gluons emitted during a grazing collision between a proton and an antiproton. (Gluons are the intermediary particles of the strong force that is thought to hold the quarks together inside the particles of the nucleus.)

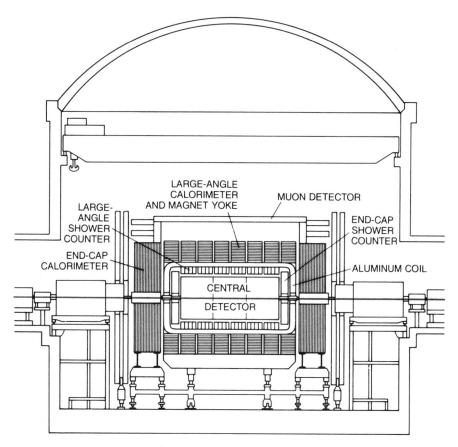

Figure 7.9 SIDE VIEW OF THE UA1 DETECTOR is shown in place in the SPS beam line. The UA1 is by far the largest detector ever built for a colliding-beam experiment: it is 10 meters long by five meters wide and weighs 2,000 tons. It was built at a cost of roughly $20 million. Experimental components for various purposes, including the search for intermediate vector bosons, are labeled.

array of closely spaced wires and a gas at low pressure. An electrically charged particle passing through the chamber ionizes molecules of the gas; the ions then drift to the wires, where they deposit their charge. From the pattern of charges appearing on many wires the trajectory of the particle can be reconstructed. The central drift chamber has its wires arranged in vertical planes; the two flanking chambers have their wires arranged in horizontal planes. Signals from particles crossing the wire planes can be processed by a computer to yield an image of the decay products on the face of a cathode-ray tube (see Figures 7.10 and 7.11).

Surrounding the three drift chambers are various other detectors. Just outside the innermost detector is a lead calorimeter, a device that measures the amount of energy deposited in it by a charged particle such as an electron. The calorimeter is enclosed in turn by a series of iron plates interleaved with scintillation counters to measure the energy of heavier particles such as pions by means of their interactions with iron atoms in the plates. Finally, on the outside of the apparatus there are several large chambers for the detection of muons that pass through both the lead and the iron plates.

Another large detector, the UA2 (see Figure 7.12) is designed more specifically to search for intermediate vector bosons. It has no magnetic field but instead relies on a large array of calorimeters similar to those in the UA1 detector to measure the energy and direction of the emerging particles. Detectors comparable to the UA1 and the UA2 are planned at

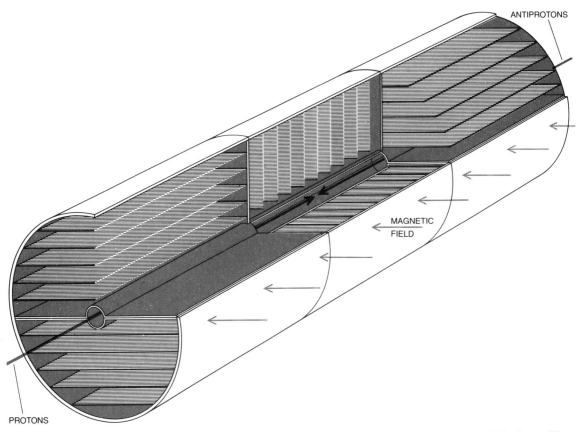

ANTIPROTONS

MAGNETIC
FIELD

PROTONS

Figure 7.10 CENTRAL DETECTION SYSTEM of the UA1 consists of three cylindrical "drift" chambers, each containing an array of closely spaced wires and a gas at low pressure. In all three chambers the wires are strung horizontally. In the central chamber the horizontal wires are arranged in vertical planes; in the two flanking chambers they are arranged in horizontal planes. A charged particle passing through the chamber ionizes molecules of the gas, which then drift to the wires, depositing their charge. The pattern of charges appearing on many wires is recorded electrically and can later be analyzed by a computer to reconstruct the trajectory of the particle on the face of a cathode-ray tube. The chambers are approximately three meters in diameter. The wires are spaced about three millimeters apart.

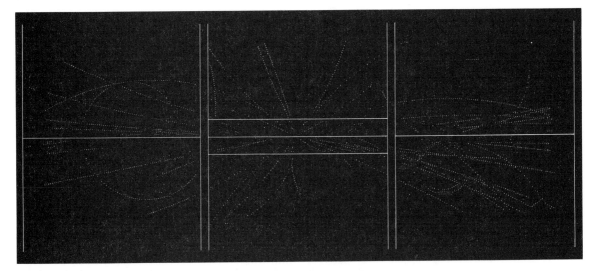

Figure 7.11 VISUAL RECORD of a proton-antiproton collision that took place in the central detection system of the UA1 was made by photographing a computer-generated display. More than 250,000 events of this type have been registered so far in the computer. Millions more will be recorded when the search for intermediate vector bosons resumes at CERN. The applied magnetic field bends the paths of the charged particles.

Figure 7.12 ANOTHER LARGE DETECTOR, the UA2, has been completed at CERN. It was designed more specifically to search for intermediate vector bosons. Unlike the UA1, it has no magnetic field. It is installed in a second experimental area, built more than 60 meters underground, some distance away from the UA1. The detector is shown in place in the SPS beam line; when it is not in operation, it is rolled back into the large work area in the foreground. The large cylindrical shaft at the top is used to lower bulky components from the surface. (Photo by European Organization for Nuclear Research.)

Fermilab. If intermediate vector bosons exist, we believe these detectors will be adequate to discover them and to investigate their properties, thereby confirming the unified electroweak theory. If Higgs bosons exist, they might also be detected, thereby providing further support for the theory. Of course, it is still possible that the electroweak theory is incorrect and that none of these particles is real. One way or the other, the answer should be known soon.

POSTSCRIPT

The charged intermediate vector boson, $W^{\pm}$, was discovered in January 1983, followed by the neutral intermediate vector boson, Z^0, several months later. These discoveries were made by international teams of scientists using the antiproton-proton collider at CERN. Their technique was exactly as we had described. Their discovery marks the culmination of nearly 50 years of searching for the origin of the weak force as described in Chapter 1 by Chris Quigg and by Gerard 't Hooft [see "Gauge Theories of the Forces between Elementary Particles"; SCIENTIFIC AMERICAN, June, 1980]. The Glashow-Salam-Weinberg (GSW) electroweak theory predicted the mass of these intermediate vector bosons. By 1986 nearly 1,000 $W^{\pm}$ and 100 Z^0 events had been accumulated and their properties and mass values were well within those of the GSW theory: $M_W^{\pm} = 82$ GeV and $M_Z^0 = 92$ GeV. The 2 TeV antiproton-proton collider at Fermilab has also detected intermediate vector bosons. The asymmetry in the charged lepton spectrum we discussed has been observed, as has the parity violation.

The enormous success of the GSW theory has led to a search for another of its predicted particles, the Higgs boson. This search will be initially carried out for "low" masses at the CERN and Fermilab hadron colliders, as well as at machines employing electron-positron collisions. If the mass of the Higgs boson is beyond 100 GeV, a much higher energy machine will be required, the Superconducting Super Collider (SSC) described in Chapter 5 by David Jackson, Maury Tigner and Stanley Wojcicki.

The success of the present picture of elementary particles rests heavily on the discoveries of the intermediate vector bosons. However, that theory is still incomplete and other, new particles and/or forces may be needed to reach the next level of understanding of elementary particle properties. For this reason it is important to continue the careful study of intermediate vector bosons as well as the search for other processes. For example, the mass of the $W^{\pm}$ and Z^0 particles should be known to a precision of a few percent during the next few years. Also, the decay properties should be comparably well measured. In addition, the new generation of electron-positron colliding machines should provide additional precise information about the Z^0 particles.

Another type of particle being searched for is the t quark. Its discovery would complete the three generations of quarks that we know to exist. In addition, fourth-generation particles are being searched for, such as the heavy lepton, whose less massive counterpart, the tau, is described in Chapter 6 by Martin Perl and William Kirk. Experimentally, we now know that the mass of that fourth-generation lepton is above 41 GeV, or more than one-half the mass of the $W^{\pm}$ and Z^0 particles.

Between 1982 and 1985 we saw an incredible advance in our understanding of elementary particles through the W and Z discoveries. The next few years should see a round of precision measurements that follow such periods of discovery. If the SSC comes into operation, there may once again be a surge of discoveries, possibly including the Higgs boson, that will complete a large part of our picture of elementary particles and their forces.

SECTION

IV

STRONG INTERACTIONS

· · ·

The Upsilon Particle

Its unexpected discovery as the heaviest particle has prompted physicists to introduce a massive new quark, raising the number of these unobserved elementary subparticles from four to five.

· · ·

Leon M. Lederman
October, 1978

The search for the ultimate, indivisible constituents of matter that began with the pre-Socratic, atomistic natural philosophers continues unabated after 2,400 years. In the past few decades the number of identified subatomic particles has risen to more than 100, as powerful machines were developed for smashing bits of matter together and studying the scattered by-products. At first physicists believed these particles could not be broken down into smaller entities. Then they found that only the four leptons (the electron, the muon and two kinds of neutrino) seemed to be truly elementary in the sense of having no measurable size and no constituent parts. The rest of the particles, the hadrons (including the proton, the neutron and the pion), turned out to be complex objects that showed signs of an inner structure. In 1964 the quark hypothesis, which has been a cornerstone of particle physics ever since, was introduced as a description of that structure. It held that the hadrons were all ensembles of only three elementary entities named quarks. An additional quark was soon postulated, for both theoretical and experimental reasons. Although none of the four quarks has ever been observed, in spite of many attempts to isolate one, there are good grounds for believing they exist.

Last year a group of investigators (of whom I was one) from Columbia University, the State University of New York at Stony Brook and the Fermi National Accelerator Laboratory (Fermilab) discovered a new particle with a mass whose energy equivalent is 9.4 GeV (billion electron volts), a mass more than three times greater than that of any subatomic entity previously identified. Designated upsilon (Y), the new particle points to the existence of a fifth quark, one more massive than any of the others. Since the original four quarks could account for all the known properties of hadrons, a fifth subparticle seems superfluous. Its existence appears to be a mixed blessing for the quark hypothesis. On the one hand it should help physicists to determine the nature of the hitherto inscrutable quark forces. On the other the very proliferation of quarks could topple the central hypothesis that they are the most fundamental constituents of matter. After all, quarks were first introduced to account for the ever increasing number of hadrons. Now it is the quarks that are growing in number, and there seems to be no theoretical reason that would block the discovery of even more massive ones.

The research that led to the discovery of the upsilon began in 1967 at the Brookhaven National Lab-

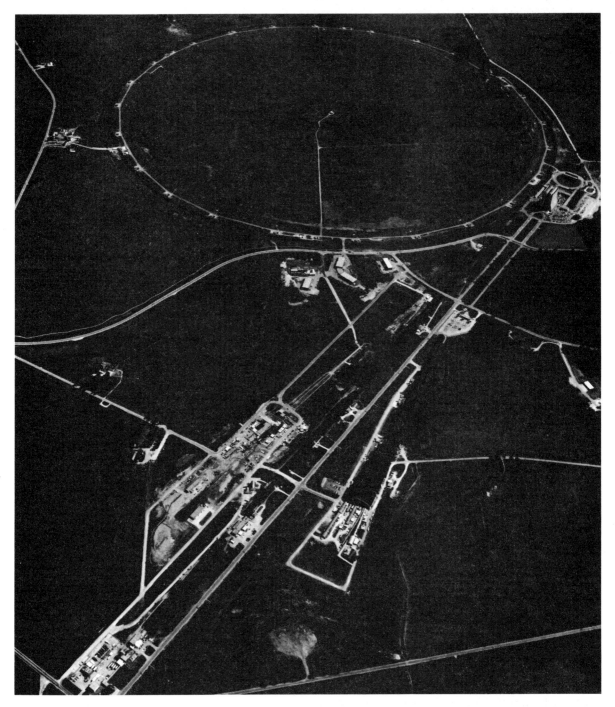

Figure 8.1 PROTON SYNCHROTRON at Fermi National Accelerator Laboratory (Fermilab) was used to generate muon pairs in experiments that led to the discovery of the upsilon particle. Here the 400-GeV (billion-electron-volt) machine appears as the large circle, which has a circumference of four miles. Tangent to the circle are long tunnels that carry particles to experimental stations. The upsilon work was done in the proton laboratory, which is in the large area at the lower left that looks as if it is under construction. Fermilab stands on a three-by-three-mile tract 30 miles west of Chicago in Batavia, Ill. Built by the Atomic Energy Commission under contract with a consortium of 53 universities, the laboratory facilities are used by groups from all over the world. Under the direction of Robert R. Wilson the accelerator, which went into full operation at 200 GeV in 1972, was upgraded to 300 GeV in 1973 and to 400 GeV in 1974. (Photo by Fermilab.)

oratory. With the 30-GeV Brookhaven synchrotron we fired energetic protons (*p*) at uranium nuclei consisting of neutrons and protons, collectively known as nucleons (N). We wanted to study what happened when a pair of oppositely charged leptons (*l*⁻ and *l*⁺) emerged, a reaction that can be written $p + N \rightarrow l^- + l^+ +$ anything. "Anything" means we had no interest in the other particles produced. Before I describe our experiments let me provide somewhat more background on leptons so that the reader will better understand why we worked so intensively with them for 10 years.

Leptons are distinguished from other subatomic particles in that they are not subject to the "strong" force that binds protons and neutrons together to form atomic nuclei. As a result energetic leptons have great power to penetrate matter. The neutrino (*v*), for one, has no electric charge and could pass through millions of miles of lead without colliding with anything. The muon (*μ*), which weighs 200 times more than the electron (*e*⁻) but otherwise exhibits identical properties, is slowed when it moves through matter by the burden of having to drag its electric charge through other electric charges. Nevertheless, because such electromagnetic forces are 100 times weaker than the strong force, the muon could penetrate many meters of iron. With a charge identical with the muon's, the electron is stopped more easily because of its smaller mass; it cannot plow its way through iron as the heavier leptons can.

T he lepton pair (*l*⁻ + *l*⁺) created in the reaction at Brookhaven had the same quantum properties as the quantum of electromagnetic energy: the photon (*γ*). This was apparent from the ease with which a photon changes into either a muon pair (*μ*⁻ + *μ*⁺) or an electron-positron pair (*e*⁻ + *e*⁺), il-

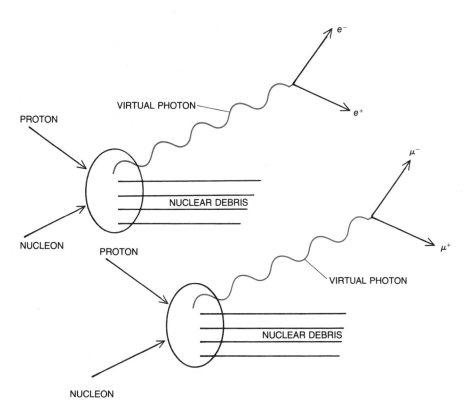

Figure 8.2 COLLISIONS BETWEEN PROTONS AND NU-CLEONS (either protons or neutrons) sometimes generate virtual photons that decay immediately into electron-positron pairs (*e*⁻*e*⁺) or into pairs of oppositely charged muons (*μ*⁻ or *μ*⁺). The bottom reaction will require much more energy than the top reaction because the muon is 200 times more massive than the electron. The additional mass of the muon, however, will enable it to penetrate much deeper into matter. The nuclear debris from these particular reactions will have very little penetrating power.

lustrated by the reactions $\gamma \to \mu^- + \mu^+$ and $\gamma \to e^- + e^+$.

A major difference between photons and lepton pairs is mass. Whereas the lepton pair has a positive rest mass when it is regarded as a single particle moving with a velocity equal to the vector sum of the motions of its two components, a photon always has zero rest mass. This difference can be glossed over, however, by treating the lepton pair as the offspring of the decay of a short-lived photonlike parent called a virtual photon. The concept of the virtual photon also appears in other reactions where the electric and magnetic properties of matter are being examined. The laws of the conservation of energy and of momentum enabled us to routinely compute the mass, energy and momentum of the virtual parent, in spite of its evanescent nature. To determine its mass (M) we had only to measure the energy of the l^- and l^+ particles emerging from the collision. The formula $M^2 = 4E^- E^+ (\sin^2\theta)$ told us that we were dealing with a massive parent whenever both the angle (θ) between the leptons and the product of their respective energies ($E^- E^+$) were large.

As long ago as 1967 we recognized in a vague and intuitive way that the emission of virtual photons could be indicative of unexplored domains inside the colliding nuclear particles. We reasoned that when an extremely energetic proton collided with a target nucleon, a highly excited and complex state would be generated. Most of the time this state would lose energy with the emission of such strongly interacting particles as pions and kaons. Occasionally, however, deexcitation would result in part from the emanation of virtual photons that would decay immediately into lepton pairs.

We had expected the masses of the virtual particles, as computed from measurements made on the leptons, to be distributed continuously (see Figure 8.3). Because we had recognized that smaller masses would be easier to create than larger ones, we thought that the yield of virtual photons would fall steeply as their mass increased. Although we did not expect the mass calculations to cluster around any particular value, we hoped this would happen. Such a cluster is called a resonance. If a resonance did manifest itself, it would indicate that the lepton pairs emanated not from some virtual entity but from some real particle. On the basis of Werner Heisenberg's uncertainty principle we could then estimate the size of whatever material within the colliding nucleons had served as the source of the new particle. Heisenberg's principle suggests that the greater the particle's mass, the smaller the size of its source. This meant that if we discovered sufficiently massive resonances, we would in fact be detecting extremely small structures within the target nucleons.

Our search for such lumps within the target nucleons was undertaken in 1967 in spite of the widespread view that matter in highly excited states was smooth and homogeneous. Moreover, even if such small but massive entities did exist, our equipment might not be sensitive enough to detect them. Other experimenters had already discovered that low-mass resonances were extremely rare; in the Brookhaven accelerator a lepton pair with a mass close to that of a proton would be created only once in a million collisions. Larger masses would be produced even less frequently, and for every one produced millions of strongly interacting particles would also be produced. Our detector would have to be capable of sorting out the rare lepton pairs from the abundant background hadrons.

After much discussion we realized we could build a detection system based on the fundamental fact that leptons can penetrate matter and hadrons cannot. Since muons can travel deeper into matter than electrons, we decided to concentrate on them and to ignore any electron-positron pairs also created. That led us to put 10 or more feet of iron between the uranium target and the lepton-pair detector (see Figure 8.4). The iron would absorb the strongly interacting particles but allow the muons to pass through and trip a series of scintillation counters.

The drawback of this detection system was that it would alter the trajectories of the muon pairs. The atoms of the iron not only would decelerate the muons, causing them to lose energy, but also would push and pull on their electric charge, deflecting them from their original paths. We were therefore in a predicament. If we measured the energy and the angle of separation of the muons after they emerged from the iron absorber and used these values to calculate the mass of their virtual parent, we would get an inaccurate answer. Yet we could not make a more accurate calculation by looking at the muons before they entered the absorber because at that point the enormous flux of hadrons would interfere with the counters. At this early stage in our work we were not too concerned about having to settle for an imprecise calculation. The goal was to detect heretofore unseen resonances at high masses,

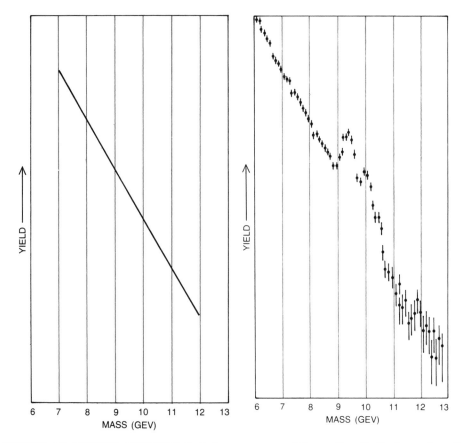

Figure 8.3 MASSES OF VIRTUAL PHOTONS that decayed into muon pairs were expected to be distributed continuously (*left*). This turned out to be the case, although in addition there was an unexpected cluster at about 9.4 GeV (*right*). Such a cluster, called a resonance, marked the presence of the upsilon. Vertical error bars through each data point represent the uncertainty as to where it should be plotted. The smaller the number of events, the larger the uncertainty.

and we believed our apparatus would do this even though it would also distort the characteristics of such resonances. That would be a small price to pay if we could discover a new particle.

We began collecting data in the fall of 1968. A digital computer processed the information and drew a graph of the yield of muon pairs observed at each mass. Since we were studying an unexplored reaction, we had no idea what the distribution would look like. Nevertheless, we were startled by the drop that began at about 1.5 GeV, flattened out just above 3 GeV and then plunged precipitously at the upper limit of our detection system, where we were not able to collect much data

(see Figure 8.5). This "shoulder" excited us. We wondered if it could represent a sharp resonance that was smeared by our crude apparatus but marked the presence of some new particle. When we lowered the energy of the bombarding protons, the shoulder would not go away. That was a good sign. It meant that the curious distribution was probably not the spurious result of some undetected quirk in the equipment. The burden of proof, however, was still on us. We could not completely dismiss the possibility that the distortion effects of the apparatus might be so overwhelming that they had spuriously warped the low-energy distribution as well. Moreover, we had to consider the possibility that the shoulder might be a peculiar characteristic

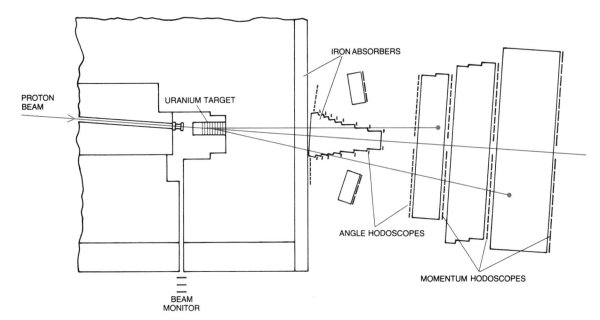

Figure 8.4 APPARATUS AT 30-GEV ACCELERATOR of Brookhaven National Laboratory generated muon pairs when protons struck uranium. The muons passed through iron that absorbed unwanted nuclear debris. Hodoscopes measured the muons' angles and momentum.

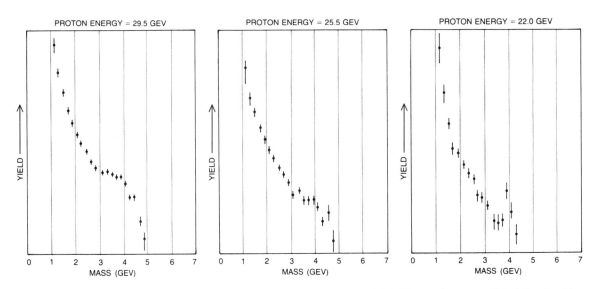

Figure 8.5 UNEXPECTED "SHOULDER" in the masses of the virtual photons generated at Brookhaven would not go away when the energy of the bombarding protons was lowered from 29.5 GeV (*left*) to 25.5 GeV (*middle*) and then to 22.0 GeV (*right*). The result suggested that the shoulder was real, perhaps the poorly resolved resonance of a new particle and not the counterfeit product of apparatus malfunction.

of the smooth distribution of virtual photons rather than the smeared resonance of a new particle.

Tentative as our uninterpreted results were, theoreticians took an immediate interest in them because they seemed to relate to the quark hypothesis. The original hypothesis of 1964 suggested that all known hadrons were composed of three quarks, labeled *u*, *d* and *s* (for "up," "down" and "strange"), and three corresponding antiquarks, $\bar{u}$, $\bar{d}$ and $\bar{s}$. Although the original quark model beautifully and simply accounted for the static properties of the more than 100 hadrons, it did not describe their dynamical properties. By 1968, however, pio-

neering workers had used the quark model to explain scattering data and collisional processes. The main difficulty with their explanation was its lack of uniqueness: reasonable alternative hypotheses that did not incorporate quarks could account for dynamical characteristics just as well.

Our lepton-pair data turned out to provide a considerable boost to the quark explanation of hadron dynamics. In 1970 two Stanford University physicists, Sidney D. Drell and Tung-Mow Yan, tried to use a quark model to generate our lepton-pair results theoretically (see Figure 8.6). Their predictions matched our data fairly well near 2 GeV but fell

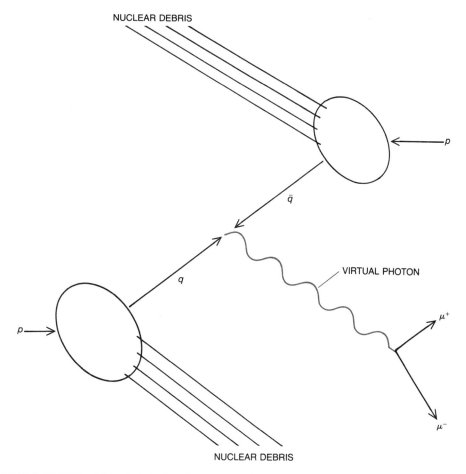

Figure 8.6 QUARK MODEL, although introduced to explain the static properties of particles, can also account for such dynamical processes as the creation of muon pairs. Sidney D. Drell and Tung-Mow Yan of Stanford University proposed that a virtual photon that decays into a muon pair is formed when a quark (*q*) from the bombarding proton and an antiquark ($\bar{q}$) from a quark "sea" associated with the target nucleon annihilate each other. Drell and Yan tried to predict the Brookhaven data, succeeding fairly well at masses near 2 GeV but not with those near 3 GeV.

below them near 3 GeV. Encouraged by this partial correlation, by the intriguing possibility of clustering and by the tremendous interest of theorists in our results, we decided to run an improved version of our experiment on the more powerful accelerator at Fermilab. The accelerator's tremendous energy, which was at that time 300 GeV, would increase the probability of pair emission at 3 GeV, and we hoped this would finally enable us to identify the significance of the mysterious shoulder.

Then in 1974, before we began taking data, the three-quark model was overthrown by what was called "the November revolution." The discovery of a new particle was independently announced by Samuel C. C. Ting of Brookhaven and the Massachusetts Institute of Technology and by Burton D. Richter of Stanford. At Brookhaven the new particle, which was named by Ting J and by Richter ψ (psi), showed up as a spectacular enhancement in the masses of virtual photons that had decayed into electron-positron pairs (see Figure 8.7).

The discovery of the J/ψ resolved several significant problems in particle physics. It explained our lepton-pair data, and it suggested the existence of a fourth quark, designated c (for "charm," the new quantum-mechanical property it implied). The shoulder we had seen in 1968 was now interpreted as being a badly smeared version of the J/ψ narrow enhancement at 3.1 GeV. The revolutionary aspects of the J/ψ lay in this very narrowness. According to Heisenberg's uncertainty principle, a narrow, or well-defined, mass implies a lifetime that is long compared with that of most other subatomic particles. And a long life span meant that the J/ψ was inhibited from decaying into such particles as pions and kaons. The existence of a fourth quark could explain why this was so. Since quarks are truly fundamental, one kind of quark cannot easily turn into another. If the J/ψ's were made up of only the charmed quarks, they could not easily decay into pions and kaons, which are made up of only the other three quarks. Subsequent investigations supported the interpretation of the J/ψ as a bound state of a fourth quark and its antiquark. The concept of a fourth, charmed quark was further confirmed when particles were discovered that seemed to consist of various combinations of all four quarks.

A comparison of our shoulder distribution of 1968 and the J/ψ data of 1974 bore out our conviction that what we had gained in sensitivity we had lost in resolution. We had detected more than 10,000 muon pairs with our highly sensitive appa-ratus, but we could not interpret the smeared distribution. The discoverers of the J/ψ at Brookhaven, on the other hand, used a new generation of particle detectors to find only 242 pairs, but because their apparatus could locate the positions of the pairs on the mass scale with greater accuracy they saw a highly resolved, narrow peak (see Figure 8.8).

Now that the mystery of the shoulder had been solved, we decided to use the new Fermilab accelerator to look for resonances in the unexplored mass range above 5 GeV. In 1975 and 1976 we observed hundreds of events in three lepton-pair runs. The energy of the Fermilab accelerator had been boosted to 400 GeV, an increase that would turn out to be crucial for our work. This time we could monitor the distorting effects of our apparatus by examining how it altered the J/ψ resonance, which we could not have done in 1968. We also had years of experience with muon pairs and of progress in detector development that we could put to good use.

In February of 1977 our group began to assemble a new version of the lepton-pair experiment utilizing what we had learned over the preceding two years. We realized that in order to draw any conclusions about the rarer, higher masses we would have to observe many more events. At the same time we would have to improve the resolution or we would be confronted with the same kind of uninterpretable data we had collected in 1968.

John Yoh of Columbia had noticed a small number of events near 9.5 GeV in our 1976 results. He put a bottle of Moët champagne, labeled "9.5," in our group's refrigerator. This convinced no one that we were on the track of a new particle. We were nonetheless encouraged in our search by the fact that our data were unique: no one else had ever seen 350 lepton pairs with masses greater than 5 GeV. There might just be something there—somewhere.

Experience showed we could move the detectors closer to the target so that more muon pairs would reach them. Stephen W. Herb of Columbia had predicted correctly that this would not also increase the enormous flux of hadrons interfering with the detection system. In 1968 we had used iron to absorb the unwanted particles, but the iron atoms, with their 26 protons and 26 electrons, exerted an electromagnetic force on the muons that deflected them from their original paths. As a result the masses of the muons could not be accurately calculated. This

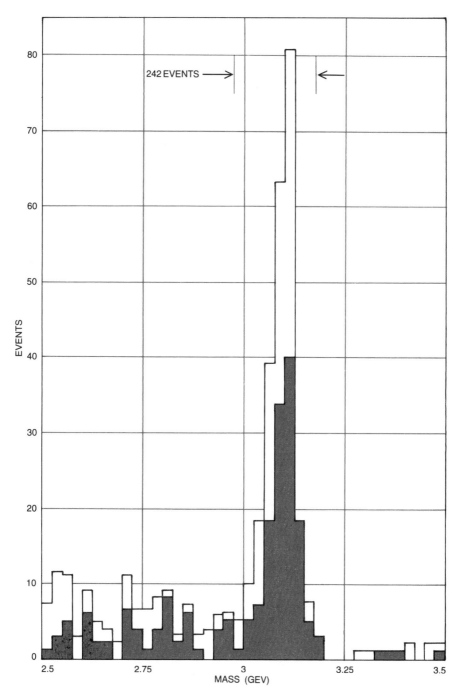

Figure 8.7 J/ψ PARTICLE was discovered in 1974 as a narrow resonance in the masses of virtual photons that decayed into electron-positron pairs. The colored distribution represents the yields of masses obtained when the detection spectrometer was run with the bombarding particles at a normal intensity; the white distribution, the yield when the intensity was cut by 10 percent. The resonance at 3.1 GeV, which showed up clearly in both runs, was interpreted as a highly resolved version of the shoulder found at Brookhaven in 1968. The J/ψ particle pointed to the existence of a fourth quark, labeled c for "charm," and a corresponding antiquark, labeled $\bar{c}$.

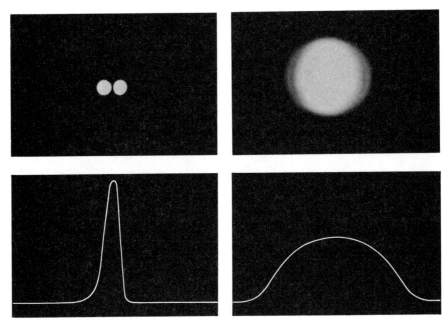

Figure 8.8 PROBLEM OF POOR RESOLUTION in high-energy experiments is illustrated by comparing photographs with curves of experimental results. An unfocused camera can make two lights (*top left*) appear to be one (*top right*). The Fermilab apparatus made the J/ψ particle's narrow resonance (*bottom left*) look broad (*bottom right*). A computer helped to clarify the distortion.

time we would use the metal beryllium as the principal absorber. With only four protons and four electrons, the beryllium would hardly deflect the muon pairs, although it would still be able to screen out most of the hadrons (see Figure 8.10).

One major obstacle remained. Muons are treacherous particles. As I have mentioned, they can easily penetrate many meters of iron. We needed absolute assurance that our muon pairs were honest: that they had been born in the target and had proceeded undeviated and unscattered through a large deflection magnet and into our counters. To gain this assurance we wanted to put a detector in the middle of the magnet. This proved difficult (it was somewhat like designing a delicate and precise watch to operate inside a blast furnace), but Walter R. Innes of Fermilab came up with a successful design. Still, we were not satisfied with the setup. The events of greatest interest were also the rarest ones. Over a long day with more than 10 billion nuclear collisions per second extremely improbable happenings could conspire to spoof the experiment. To guard against this possibility Charles N. Brown of Fermilab designed a simple magnetic system that would remeasure each muon's energy after it emerged from the main detector.

On May 1 of 1977 we gathered our first data. We were elated to find that our improved apparatus registered 90 times more muon pairs than it had the year before. The upgraded accelerator had functioned superbly, supplying unlimited quantities of protons with needle-sharp precision. In the first week we observed 3,000 muon pairs with energies higher than 5 GeV, more than 10 times the rest of the world's data and of much better quality. We graphed the results, and they seemed remarkably free of the interfering effects of hadrons. The J/ψ resonance showed up clearly, which meant we had succeeded in increasing the resolving power of our apparatus. Our excitement rose to a high pitch when we saw that the steady decrease in the yield of muon pairs, as they became more massive, was interrupted near 10 GeV by an intriguing bump.

The following week we doubled our data and still the bump remained. Although we could no longer dismiss it as a misleading happenstance, we wondered if it could be the wayward product of some undetected idiosyncrasy of our apparatus. Perhaps the deflecting magnets or the counters had malfunctioned. Fortunately we had mechanisms for checking that out. We looked separately at each square centimeter of the detector's surface to see how the muons that struck each area were distributed. Everywhere we found smooth distributions, indicating that the apparatus had not generated the resonance. Moreover, when we artificially mixed Monday's μ^+'s with Tuesday's μ^-'s to form fake $\mu^+ \mu^-$ pairs, we got a perfectly smooth distribution that conformed to all our expectations about how the equipment worked. As the apparatus passed other tests and as we accumulated more data we became convinced that the resonance represented something real: a new particle with a mass of 10 GeV. Although we wanted to keep our results secret until we could fully interpret them, rumors of our discovery spread rapidly throughout the physics community. Therefore on June 20 we made our data

public: 26,000 pairs, almost 100 times the data of all previous experiments combined. We named the particle upsilon.

We next set out to determine the width of the resonance, using the same method by which the span of the J/ψ resonance had been calculated. In effect the width is the uncertainty in the mass of the resonance, and Heisenberg's principle associates a narrow peak (a small uncertainty) with a long lifetime, and a broad peak (a large uncertainty) with a short life span. After we had gathered more data we found that the resonance consisted of two closely spaced peaks (with a suggestion of a third) 600 MeV (million electron volts) apart and each peak 500 MeV wide. This indicated that the upsilon could exist in two and perhaps three states of slightly different energies. Before we concluded from these width values that the resonance of the

Figure 8.9 FERMILAB APPARATUS detected the upsilon. The bombarding protons and the target nucleons collided at a point just out of sight in the foreground. The electromagnets, located at the left and the right in front of the man in the foreground, deflected the muon pairs so that their energies and separation angles could be measured. The two components of each muon pair traveled through different arms of the detectors. Six feet wide, six feet high and 100 feet long, the two arms extend from the electromagnets to where the man in the rear is standing. (Photo by Fermilab.)

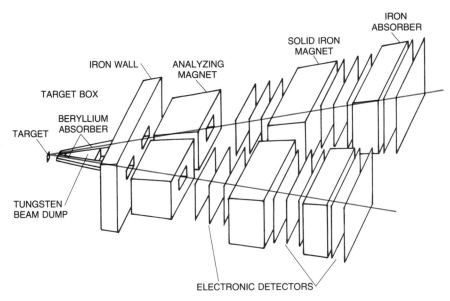

Figure 8.10 BERYLLIUM ABSORBER was used to screen out nuclear debris in the Fermilab experiment because beryllium affects the trajectories of muons much less than iron does. The tungsten beam dump collected bombarding protons that missed the target nucleons. This schematic diagram is a side view of the apparatus seen head on in Figure 8.9.

upsilon was intrinsically narrow we needed to take into account the distortion effects of the inevitably imperfect apparatus. Apparatus with a low resolving power will make nature's peaks look broad, much as a camera lens of poor quality will blur the fine details in a photograph (see Figure 8.8). In our Brookhaven experiment of 1968 poor resolution had distorted the data to the point where they were uninterpretable.

To determine how the Fermilab apparatus had deformed the shape of the resonance we relied on a game-theory approach in which a computer simulated our entire experiment. Such simulations, called the Monte Carlo method, are ubiquitous in high-energy physics. In our case the computer, programmed to know the location and function of each piece of our apparatus, selected a configuration of two muons and traced their trajectories to the final detector. If the computer program called for the muons to encounter an absorber consisting of, say, beryllium, the program would call for the muons to be scattered just as if the muons and the beryllium were real. The computer we used was a powerful one, so that the simulation could trace tens of thousands of muon pairs through the apparatus.

We then graphed the mass distribution of the Monte Carlo events and discovered that the simulated upsilon resonance was much narrower than the measured one. This suggested that the measured width was produced mainly by the apparatus. Complex computer programs do, however, have bugs. Had we caught them all? Perhaps the simulation was wrong and the resonance was actually as broad as the one we had measured. Fortunately we had ways of eliminating that possibility. Since we already knew how our apparatus had distorted configurations such as the J/ψ resonance, we could test the Monte Carlo program to see whether it correctly revealed such distortions. Indeed it did, and we confidently concluded that the width of the upsilon resonance was less than 100 MeV. This extremely narrow width indicated that our new particle had a very long lifetime.

One would normally expect that a particle with a mass as great as 10 times the mass of the proton should have an enormous number of lower mass states into which it could decay, each state contributing to a shorter lifetime. But contrary to expectations the upsilon, the heaviest particle ever discovered, had a long lifetime. This means that it does not decay into the less massive hadrons, all of which are composed of u, d, s and c quarks. At the time of our discovery last year, however, the known laws of physics could not explain why this was so. The conclusion was clear and exciting: some new law of physics forbids (or, more precisely, inhibits) the upsilon from decaying into ordinary hadrons.

In search of this new law we looked to see if any work in theoretical physics had anticipated our discovery of the upsilon. Over the years many theorists have suggested the existence of new particles to account for puzzling data, and we wondered if the upsilon could be one of those that had been proposed.

The only reasonable candidate was a new, massive quark bound to its antiquark in atomlike configurations that would show up as a closely spaced set of masses having many features of the upsilon. The theoretical papers that raised the possibility of such a quark were speculative discussions appealing to aesthetic prejudices. One group of papers hoped the existence of a new quark would be able to account for some curious results of certain neutrino-scattering experiments. Our best calculations, made early this year, indicate that the upsilon has resonances at 9.4, 10.0 and 10.4 GeV (see Figure 8.11). A particle made up of a fifth quark and its antiquark might exist in a ground state, or lowest state, at 9.4 GeV and in excited states at 10.0 and 10.4 GeV. Moreover, the existence of a fifth subparticle would neatly account for the long lifetime of the upsilon, just as the fourth quark had accounted for the long life span of the J/ψ. If the upsilon consisted only of the fifth kind of quark, it could not decay into ordinary hadrons, which consist of various combinations of only the other four quarks. Such compelling considerations convinced most particle physicists that the upsilon is indeed an atomlike composite of a fifth kind of quark bound to its antiquark.

An amusing thing about the hypothesis of a fifth quark is that the reasons for which some theoretical papers introduced it turned out later to be specious. The new subparticle was supposed to explain puzzling data from scattering experiments, which on closer examination proved not to be puzzling at all. Where the fourth quark had accounted for all kinds of enigmatic phenomena, the fifth quark explained only the results of our experiment. The misinterpreted scattering data had nonetheless served a valuable heuristic purpose in stimulating specula-

tion about the properties of particles composed of heavier quarks, properties the upsilon turned out to have. The fact that the four identified quarks were paired off as "up" and "down" and as "strange" and "charm" had led theorists to predict that if there were a fifth quark, there would also be a sixth. The eccentric names "top" and "bottom" or "truth" and "beauty" had been reserved for the two new quarks in the event that they were discovered.

The upsilon resonances present physics with an embarrassment of riches: an unexpected family of new particles composed of an unexpected fifth quark. The impact of the upsilon has already been far-reaching. It has prompted searches for other heavy particles in hitherto unexplored ranges of mass, and it has shed light on the inscrutable strong force. This force, which binds quarks together into hadrons and hadrons together into atomic nuclei, is too powerful to investigate by conventional scattering and collision techniques. Yet any proposed model of the strong force, being a description of the force between a quark and an antiquark, should correctly predict the energy (or mass) levels of the upsilon family. With the upsilon, as opposed to the hadrons of lower mass, it is easier to evaluate these predictions because the velocities of massive quarks are comparatively low. This means that complicated relativistic considerations never enter the calculations. The predictions of several such theoretical models have failed, which removes them from consideration. The successful models all suggest that the new fifth quark has a charge of $-1/3$ (the charge of the electron being -1) and that the force between quarks increases with the distance between them. Such a force had been proposed to account for the failure of particle physicists to observe quarks in the free state: so much energy is needed to increase their separation that when energy is supplied, it goes into creating new quark pairs rather than into splitting old ones. The current thinking is that quarks may be permanently confined to composite structures.

The confirmation of this suggestion and a better understanding of the strong force may not come, however, from the study of upsilons created in proton accelerators, such as the ones with which we worked at Brookhaven and Fermilab. The production process is too complex. For example, the initial collision at Fermilab involves three quarks in the proton projectile smashing into three quarks in the target nucleon. The next step in the demystification of quark forces will come when physicists are able

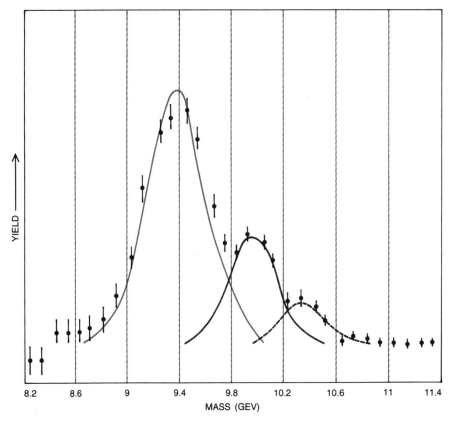

Figure 8.11 THREE CLOSELY SPACED RESONANCES characterize the upsilon. With a lowest state at 9.4 GeV and excited states at 10.0 and 10.4 GeV, the particle was interpreted as consisting of a massive fifth quark bound to its antiquark. The upsilon's spectroscopy will be clarified by experiments with storage-ring accelerators, and the search for a sixth quark has already begun.

to intensively study upsilons that have been produced in storage-ring machines in which electrons and positrons circulate in opposite directions and collide with one another. These machines will provide cleaner and more highly resolved data. Indeed, they already have. In April a group of workers at the DESY (for Deutsches Elektronen Synchrotron) laboratory in Hamburg modified their electron-positron storage ring to look for the upsilon. They found it at 9.46 GeV and were able to place an upper bound of 7 MeV on the width of its resonance. (This is a substantial improvement on our value of 100 MeV.) Their data also suggest a −1/3 charge. The spectroscopy of the upsilon system will surely be worked out in detail over the next few years when powerful new storage rings are completed at Hamburg, Stanford and Cornell University. Physicists will be looking for the sixth quark, the expected mate of the fifth, and they will be searching for particles made up of combinations of the six quarks.

As accelerator techniques advance, physicists will undoubtedly continue to discover new subatomic entities. The proliferation will raise deep, unsettling questions. Are the kinds of quark limited in number? If there are six, why not 12? If there are 12, why not 24? And if the number of kinds of quark is large, does it make sense to call the quarks elementary? The history of science suggests that the proliferation of physical entities is a sign the entities are not elementary. The chemists of the 19th century reduced the apparently infinite variety of chemical substances to some 36 elements, which escalated over the years to more than 100. As indivisible, ultimate constituents of matter the chemical elements simply proved to be too many. In the 1930's it was discovered that all the elements were made up of electrons, protons and neutrons. After World War II these particles were joined by dozens of others: pions, kaons, lambda particles and so on. Again there were too many. Then it seemed that all of these could be reduced to three quarks. Now experiments indicate that a fourth and a fifth quark exist. Are they also too many? Will simpler structures from which quarks are made soon be proposed? Is it possible that there are no elementary particles at all, that every entity in nature has constituent parts? Or will the ultimate simplicity that most physicists believe in be lodged in the mathematical groups that order the particles rather than in truly elementary objects?

Putting aside these last highly speculative ideas, most physicists despair of addressing such questions because of the difficulty, if not impossibility, of examining a quark in isolation. Yet the experience with the upsilon particle indicates that in spite of this difficulty detailed knowledge of the motions and forces of quarks can be acquired. The apparent inseparability of these entities should not in itself block the path of inquiry. Consider the lesson physicists should have learned from the electron. The developmnt of the theory of the electron would probably have been slowed but not otherwise hampered if electrons had only been observed bound in atoms and never in the free state. This is an experimentalist's response to the prophets of gloom who view the confinement of quarks as an ultimate limitation on knowledge: a wall erected by nature to hide its last secrets forever. And who is to say that physicists will never build an ultrapowerful accelerator that could overcome the confining force and liberate the quark?

EDITOR'S NOTE

The author wishes to acknowledge the contribution to the work described in this article made by the following individuals: J. A. Appel, B. C. Brown, C. N. Brown, J. H. Christenson, S. W. Herb, G. S. Hicks, D. C. Hom, W. R. Innes, A. S. Ito, H. Jöstlein, D. M. Kaplan, R. D. Kephart, P. J. Limon, B. G. Pope, J. C. Sens, H. D. Snyder, K. Ueno, T. Yamanouchi, J. K. Yoh and E. Zavattini.

POSTSCRIPT

The discovery of the upsilon at Fermilab in three narrow states with masses near 10 GeV in 1977 established the existence of the b (for "bottom" or "beauty") quark. The key to the interpretation of the Fermilab experiment was the notion that the "upsilon" is a bound state of a new quark, b, and its antiquark twin, $\bar{b}$. The three states are then identified as the ground state and the first two excited states of the two-quark system. In the notation of the hydrogen atom those states are the $n = 1, 2$ and 3 radial quantum states of this unique atom.

The confirmation of the quark-antiquark (quarkonia) interpretation of these dimuon peaks came soon in observation of positron-electron collisions at the DESY storage ring system in Germany. Together with the discovery of the tau lepton at SLAC, described in Chapter 6 by Martin Perl and William Kirk, a third quark-lepton generation was unveiled and with it the symmetry suggesting that quarks and leptons were assembled into generations whose total electric charge is zero.

Gaining information about the quark-quark force has been one of the most fruitful applications of the heavy quarkonia states. In the hydrogen atom the electron is bound to the proton by an electrical force described by Coulomb's law. In the upsilon "atom," the force between the b and the $\bar{b}$ is the primordial strong force which we very much want to study. A nonrelativistic treatment of these massive particles is accurate, making the upsilon, a beauty-antibeauty state, simpler than the charm-anticharm states of the J/ψ family. Thus, the range of possible potential models for quarkonia was greatly narrowed by precise measurements of the beauty-antibeauty states. Indeed, many of them have since been observed in both the S-state excitations, that is, those with no angular momentum, with up to six excited levels, and in P-state excitations with one unit of angular momentum. The description of the potential well in which the quark-antiquark exists is relevant to the very deepest aspects of the problem of quark confinement.

The discovery of the upsilon opened up another field of quark physics, the study of "naked bottom" states consisting of a u or d quark bound to a b quark as described in Chapter 10 by Nariman Mistry, Ronald Poling and Edward Thorndike. Many of these electrically neutral B mesons have been observed and their decay modes measured, thus providing another road for the study of the weak interaction. A major target of the late 1980's is to observe time reversal or charge conjugation-parity (CP) violation and mixing in the $B°\bar{B}°$ system in analogy to the famous $K°\bar{K}°$ system. Here a b quark replaces the s quark. Mixing, a necessary precursor to CP violation, may have already been observed.

Upsilons have been invoked to provide sources of axions and for hiding a 10 GeV Higgs particle. The upsilon is also a popular decay product of exotic heavier objects.

Finally, a vast literature reports on the production of states carrying the beauty-quark in collisions of hadrons whose valence quarks are u and d. Since the colliding objects are hadrons composed of the garden variety up and down quarks, how is it possible for the collisions to yield up b quarks? The answer lies in the complexities of the quark binding effects in the colliding protons and mesons. The quark forces are carried by gluons, which are rapidly exchanged in the hadron. Gluons can, in their virtual state, dissociate into a quark and an antiquark, the required mass being "borrowed" for a short time, in harmony with the Heisenberg uncertainty condition. Thus, the hadron is considered to be composed of u and d valence quarks and a "sea" of gluons and heavier (e.g., beauty) quark-antiquark pairs. A fortuitous collision can then free a b quark which is required to pick up a $\bar{d}$, say, on its way out of the collision volume.

For these collisions the gluon content is a crucial factor in production processes. This, in turn, provides another handle on the strong forces as described by quantum chromodynamics and discussed in Chapter 3 by Claudio Rebbi. Production of upsilons involves the direct observation of heavy antiquarks in the sea of the target particle and is an additional, sensitive probe of the complex parton description of nucleons.

As so often happens in science, a vast array of data from positron-electron and hadron machines now provide stringent constraints on the weak, electromagnetic and strong forces at the smallest distances, as characterized by the inverse of the quark mass. This has all emerged from the discovery of the upsilon only a decade ago.

Quarkonium

An "atom" made up of a heavy quark and an antiquark provides the best available system for examining the forces that bind together the elementary constituents of subnuclear particles.

. . .

Elliot D. Bloom and Gary J. Feldman
May, 1982

The most convenient context for investigating the forces of nature is a system of two objects bound together by mutual attraction. The earth and the moon, for example, constitute the most readily accessible system in which to observe the gravitational force. The hydrogen atom, consisting of an electron and a proton, has long been an essential testing ground for theories of the electromagnetic force. The deuteron, made up of a proton and a neutron, represents a model system for studies of the forces in the atomic nucleus. Now there is a bound system in which to investigate the force that acts between quarks, the constituents of protons, neutrons and many related particles. The system is called quarkonium, and it consists of a heavy quark bound to an equally massive antiquark. The force at work in quarkonium is the strongest one known; it has come to be called the color force, and it is now thought to be the basis of all nuclear forces.

Of the various two-body systems the simplest in some respects is the artificial atom called positronium. It is made up of an electron bound to a positron (the antiparticle of the electron). Like the hydrogen atom, positronium is held together by the attraction of opposite electric charges, but it is a more symmetric structure. Whereas the proton of the hydrogen atom is larger and heavier than the electron, the two component particles of positronium are identical in size and mass. Several states of positronium have been observed; in each state the electron and the positron have a unique mode of motion, and as a result each state has a distinctive energy. From the spectrum of energy states it is possible to deduce certain characteristics of the electromagnetic force.

Quarkonium is closely analogous in structure to positronium. In quarkonium the bound quark and the antiquark are again identical in size and mass, and the allowed modes of motion in quarkonium are similar to those in positronium, so that an equivalent spectrum of energy states can be expected. In the past several years a major effort to detect the quarkonium states has been undertaken, and about a dozen have been found. In a qualitative sense they correspond exactly to the positronium states. On the other hand, a quarkonium system is smaller than a positronium system by a factor of 100,000, and its total mass or energy is larger by a factor of from 3,000 to 10,000. Moreover, from the details of the quarkonium spectrum it is apparent that the color force is not only stronger than electromagnetism but also more complex.

Quarks were introduced into physics in 1964 by Murray Gell-Mann and George Zweig, both of the California Institute of Technology. The aim of the quark model was to explain the diversity of the particles designated hadrons, which up to then had been considered elementary. Quarks can combine in two ways to form a hadron. Three quarks bound together make up a baryon, a member of the subclass of hadrons that includes the proton and the neutron. A bound system of a quark and an antiquark constitutes a meson. The lightest and commonest of the mesons is the pion; the quarkonium systems we shall describe are also mesons, but they are much heavier.

Originally there were supposed to be just three kinds of quark, designated up, down and strange, or u, d and s. All the hadrons known could be interpreted as combinations of these three quarks (and the antiquarks $\bar{u}$, $\bar{d}$ and $\bar{s}$). For example, the quark composite uud has all the properties of the proton, and the quark-antiquark aggregate $u\bar{d}$ can be identified as the positively charged pion. It might seem that this rich variety of hadrons would offer ample opportunities for exploring the force between quarks. Actually experiments with protons, neutrons, pions and other "ordinary" hadrons can yield only indirect information about the interquark force. The reason is that the u, d and s quarks are quite light; indeed, their mass, when it is expressed in energy units, is comparable to the binding energy that holds the quarks together in the hadron. As a result the quarks in an ordinary hadron move with a speed close to the speed of light, and calculations of their properties must be done with the complicated methods of the special theory of relativity. In general such calculations are too difficult to be practical.

What was needed was a bound system of heavier quarks, in which the binding energy would be small compared with the quark mass. The quarks would then move much slower than the speed of light, and the complications of the theory of relativity could be ignored. Such a nonrelativistic quark system was found in 1974 with the discovery of an extraordinary meson having a mass (in energy units) of 3,095 million electron volts (MeV). The meson was found almost simultaneously by two groups of experimenters. One group, at the Brookhaven National Laboratory, named the new particle J. The other

group, at the Stanford Linear Accelerator Center (SLAC), called it ψ (the Greek letter psi), and that is the name we shall use here.

What is the ψ meson? It seemed unlikely to be any combination of u, d or s quarks, in part because all the combinations with appropriate properties were already listed among the known hadrons. Some 10 years earlier, however, James D. Bjorken, who was then at SLAC, and Sheldon Lee Glashow of Harvard University had speculated that there might be a fourth quark flavor, which they had fancifully named charm (c). In 1970 Glashow and his colleagues John Iliopoulos and Luciano Maiani had argued on theoretical grounds that the charmed quark must exist and should be substantially heavier than the other quarks. Not long before the discovery of the ψ Thomas W. Applequist and H. David Politzer, who were then also at Harvard, had pointed out that a charmed quark and a charmed antiquark might form a nonrelativistic bound state. They had named the bound state charmonium, in analogy to positronium. The ψ was soon recognized as a form of charmonium, that is, a meson with the quark composition $c\bar{c}$.

The discovery of charmonium stimulated a search for still heavier quarks. There was reason to think they would come in pairs, and the first two quarks after the known ones were designated bottom (b) and top (t). In 1976 Estia Eichten and Kurt Gottfried of Cornell University suggested that bottomonium (the meson with the quark constitution $b\bar{b}$) and toponium ($t\bar{t}$) should form nonrelativistic systems similar to charmonium but with a considerably richer spectrum of bound states. The first bottomonium state was discovered in 1977 at the Fermi National Accelerator Laboratory (Fermilab) near Chicago; it is called the Υ (the Greek letter upsilon), and it has a mass of 9,460 MeV. The toponium system has not been detected. If it exists, its mass must be greater than 36,000 MeV.

In the past eight years the charmonium spectrum has been surveyed in detail and a start has been made on the exploration of the bottomonium states. In addition a number of particles made up of a charmed quark in combination with a quark of another kind have been identified; such particles are said to exhibit naked charm. Recently indications of naked bottom have also been reported. Here we shall be concerned primarily with the systems in which a heavy quark is bound to an antiquark of the same kind, as in $c\bar{c}$, $b\bar{b}$ and the conjectured $t\bar{t}$. It is to these three systems that we apply the generic term quarkonium.

I n order to understand the forces at work in quarkonium it is instructive to begin with a discussion of the hydrogen atom and positronium. The chief force binding an electron to a proton or to a positron is described by Coulomb's law. The law states that the force is directly proportional to the product of the electric charges of the particles and inversely proportional to the square of the distance between them. The constant of proportionality, α, is a measure of the inherent strength of the electromagnetic force; its numerical value is approximately 1/137.

Classical physics predicts that the particles in positronium or the hydrogen atom should fall toward each other until they collide. The prediction disagrees with experiments and was troublesome in early attempts to formulate a theory of the atom; it was resolved only with the development of quantum mechanics in the 1920's. An important step in this development was the model of the hydrogen atom put forward by Niels Bohr in 1913. Bohr simply postulated that the electron in a hydrogen atom can occupy only certain discrete orbits. In each allowed orbit the electron has a definite energy, and it can change its energy only by making an abrupt

jump to another allowed orbit. In going from one orbit to another the electron either emits or absorbs a photon, or quantum of electromagnetic radiation, with an energy exactly equal to the difference in energy between the two orbits.

Bohr showed that the binding energy of the electron in a given orbit is equal to E_0/n^2, where E_0 is the binding energy of the smallest orbit and n is a positive integer called the principal quantum number. The binding energy is the energy that must be supplied to separate the electron from the proton completely. For the hydrogen atom E_0 is about 13.6 electron volts. Values of n are assigned to the orbits in sequence, beginning with the smallest orbit. For the smallest orbit n is 1 and the binding energy is 13.6 electron volts; for the next orbit n is 2 and the binding energy is $13.6/2^2$, or 3.4, electron volts, and so on.

The Bohr model can be applied to positronium, but because of the difference in mass between the proton and the positron the radius of a given positronium orbit is twice the radius of the hydrogen orbit with the same principal quantum number. The binding energy of positronium is therefore half as

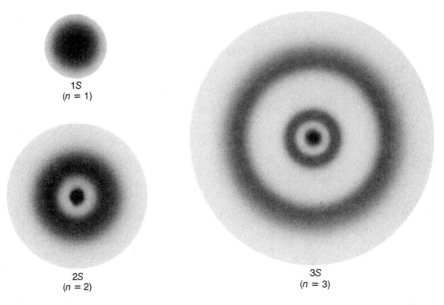

1S
(n = 1)

2S
(n = 2)

3S
(n = 3)

Figure 9.1 STATES OF A BOUND SYSTEM made up of a particle and an antiparticle are described by wave functions, which give the probability of finding the particle or the antiparticle at any point in space. The probability-density distribution is given for three states of positronium, the bound system consisting of an electron and a positron. Darker regions signify higher probability. Each state is identified by its principal quantum number, n, and by a letter code that indicates the orbital angular momentum of the system. The three configurations shown are S states, which have zero orbital angular momentum. Quarkonium forms analogous states.

great: 6.8 electron volts for the smallest orbit. The sequence of states is the same, but all energies are reduced by a factor of two.

For a complete description of a bound system it is necessary to specify not only the energy but also the angular momentum (see Figure 9.2). In a two-body system there are three contributions to the total angular momentum: the spinning of each body on its internal axis and the orbital motion of the two bodies around their center of mass. Each of these quantities can be represented by a vector oriented along the axis of rotation and having a length determined by the magnitude of the angular momentum. One could describe the angular momentum of a system by giving the magnitude and the direction of the three vectors. Other methods of representation,

however, convey the same information and better reflect the symmetries of the system. In describing the states of positronium and quarkonium it is more illuminating to specify the orbital angular momentum, the sum of the two spin angular momenta and the total angular momentum.

In a large-scale system where the laws of quantum mechanics can be neglected the angular-momentum vectors can have any magnitude and orientation. In some respects the description of angular momentum in a quantum-mechanical system is simpler. Quantum mechanics does not allow an observer complete knowledge of an angular-momentum vector; one can measure only the absolute magnitude of the vector and its projection along one axis in space (although any axis may be chosen).

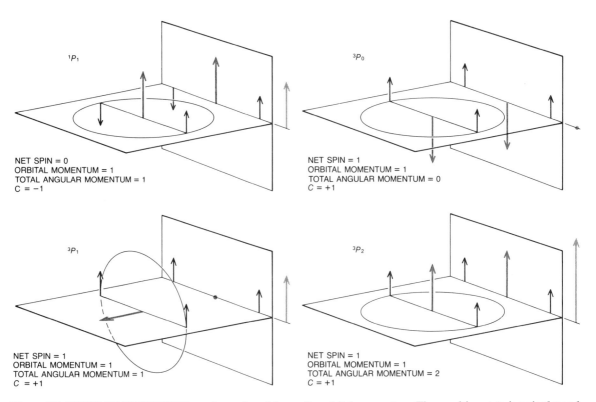

Figure 9.2 ANGULAR MOMENTUM can be analyzed by means of vectors projected onto a plane. It has contributions from the spin of the particles and their orbital motion. Both electrons and quarks have an intrinsic spin of 1/2 (*black arrows*); in the states shown the orbital momentum is equal to 1 (*colored arrows*). If the projections of the spin vectors point in opposite directions, the net spin is zero and the total angular momentum (*gray arrows*) equals

the orbital momentum. The resulting state is a singlet and is designated $^1P_1$. If the spin vectors are parallel, the net spin of 1 can have three possible orientations, designated $^3P_0$, $^3P_1$, and $^3P_2$. The charge-conjugation number (C) is determined by adding the absolute magnitudes of net spin and orbital angular momentum. (Designations are explained in text.)

Furthermore, the measured quantities can have only certain discrete values: an angular momentum must be either an integer or a half integer when it is expressed in units of the smallest possible quantity of angular momentum (which is equal to Planck's constant divided by 2π). A final constraint is that the orbital angular momentum must always be less than n, the principal quantum number.

Orbital angular momentum can take on values given by the integers in the series 0, 1, 2, 3 and so on. In a compact notation introduced to describe atomic spectra some years before Bohr developed his model, the successive values of orbital angular momentum are designed by the letters S, P, D and F. The Bohr orbit and the orbital angular momentum of a state can therefore be specified by appending the appropriate letter to the principal quantum number. The lowest-energy state of positronium, for example, is designated $1S$; it is the state with n equal to 1 and no orbital angular momentum. The symbol $2P$ designates a state with n equal to 2 and one unit of orbital angular momentum.

In both positronium and quarkonium the individual constituent particles have an intrinsic spin angular momentum of $1/2$ unit. The spins can combine in just two ways. If they point in opposite directions, they cancel each other and the system has a net spin of zero. If they are parallel, they add and the net spin is 1. The total angular momentum of the system depends on the relative orientation of the orbital angular momentum and the net spin; to be more precise, it depends on the projection of the net spin vector along the axis of the orbital vector. When the net spin is zero, it can clearly have only one possible projection, namely zero. Such a state is called a singlet, since for a given value of orbital angular momentum there is just one possible value of total angular momentum. When the net spin is 1, it has three possible projections along the orbital axis: $+1$, 0 and -1. These values correspond to the three possible orientations parallel, perpendicular and antiparallel. A state of this kind is called a triplet, since for a given value of orbital angular momentum there are three distinguishable states that differ in total angular momentum. (See Figure 9.2.)

The net spin and the total angular momentum are incorporated into the notation for quantum states by a system of superscripts and subscripts. For singlet states a superscript 1 is prefixed to the letter designation of the state; for triplet states the super-script is a 3. The total angular momentum is given by a subscript appended to the symbol. Thus 2^1P_1 designates the state of positronium with n equal to 2, one unit of orbital angular momentum and a net spin of zero. Since there is no net spin, the state is a singlet and the total angular momentum is necessarily equal to the orbital momentum. The triplet of states 2^3P_0, 2^3P_1 and 2^3P_2 have the same principal quantum number and the same orbital angular momentum, and they all have a net spin with a magnitude of 1. They are distinguished by the orientation of the net spin with respect to the orbital axis, giving rise to the three values of total angular momentum.

The Coulomb force between an electron and a positron depends only on the charges of the particles and the distance between them; it is independent of their angular momentum. If the Coulomb force were the only one at work in positronium, all angular-momentum states with the same value of n would have the same energy. Actually there are other forces. Both the spinning and the orbital motion of the electric charges give rise to magnetic fields, which in turn cause attractions and repulsions that can alter the energy of a state. For example, in the $1\ ^1S_0$ state the two spins are antiparallel and the resulting magnetic interaction is an attractive one. In the $1\ ^3S_1$ state the spins are parallel and the magnetic interaction is repulsive. As a result the former state has a slightly lower energy than the latter one.

In positronium the energy shifts associated with the angular momentum of the constituent particles are exceedingly small, at most a five-thousandth of the binding energy. Analogous effects can be observed in quarkonium, where the spin and the orbital motion of the quarks give rise to color magnetic fields. The energy differences in quarkonium are much larger, however. The overall binding energy of quarkonium is about 100 million times the binding energy of positronium, and the angular-momentum states are typically separated by a fifth of the binding energy.

The analogy between positronium and quarkonium can be extended to one additional phenomenon: the decay of the bound systems. When an electron and a positron come together, they annihilate each other, and the energy equivalent of their mass appears in the form of electromagnetic radiation. A quark and an antiquark of the same kind can annihilate each other in a similar way, although the energy initially takes the form of radiation associated with the color force. For the description of these events, however, the Bohr model of the atom is not adequate.

The Bohr model has been supplanted, of course, by a more refined version of quantum mechanics. In the modern view the electron and the positron do not have a definite orbit and indeed cannot be assigned a definite position at any given instant. Instead one can calculate only the probability of finding a particle at a given point in space. The probability is defined by a mathematical function called a wave function, which is different for each energy state. For the S states of positronium the rate of decay by mutual annihilation is proportional to the probability of finding the two particles at the same point.

The decay is governed by a number of conservation laws. For example, the total electric charge of all the particles after the decay must be the same as it was before the decay. Energy and linear momentum must also be conserved. The energy of the positronium system is roughly 1 MeV; the linear momentum, since the particles can be thought of as moving in opposite directions when they collide, is zero. The photon can carry any quantity of energy, but because its mass is zero its linear momentum is invariably equal to its energy. If positronium were to decay to yield a single photon, energy and linear momentum could not both be conserved. For this reason the decay always gives rise to at least two photons. If the two photons are emitted in opposite directions, each one can carry half of the energy of the system and their linear momenta cancel.

A property of a wave function called the charge-conjugation number leads to a further constraint on the number of photons emitted. Charge conjugation is an imaginary operation in which all particles are converted into their antiparticles. When the operation is applied to positronium, the electron becomes a positron and the positron becomes an electron. Hence the system is still an atom of positronium, but the constituent particles have exchanged identities. In some states of positronium such an exchange is of no consequence; in other states, however, it changes the sign of the wave function. There is a method for keeping track of how a given state responds to charge conjugation. When the charges are interchanged, the wave function of the state is multiplied by the charge-conjugation quantum number, which has a value of +1 for the states that remain unaltered and −1 for the others. The charge-conju-

gation number, or C, is conserved in all interactions mediated by either the electromagnetic force or the color force.

The photon, which is its own antiparticle, has an intrinsic charge-conjugation number of −1. For a system of several photons or other particles the values of C for all the particles are multiplied to yield the total charge-conjugation number. It follows that a state of positronium with C equal to +1 must decay to yield an even number of photons; the actual number is almost always two. A positronium state with C equal to −1 must yield an odd number of photons. As we showed above, the decay into a single photon is forbidden by the conservation of energy and momentum, and so the minimum odd number is three.

What determines the charge-conjugation number of a positronium state? It can be determined by adding the absolute magnitudes of the orbital angular momentum and the net spin. If the sum of the absolute magnitudes is an even number, C is +1; if the sum is odd, C is −1. Thus for the 2^1P_1 state, with one unit of orbital angular momentum and no net spin, the sum is odd and C is −1. In comparison the 2^3P_0, 2^3P_1 and 2^3P_2 states all have one unit of orbital angular momentum and one unit of net spin, yielding an even sum and a charge-conjugation number of +1. (See Figure 9.2.)

The various properties of the wave function of a positronium state influence the lifetime of the state. For an S state the intensity of the wave function at the point corresponding to an interparticle separation of zero determines the probability of the particles' colliding, which is a precondition to their annihilation. For states whose wave functions have the same intensity at zero separation, such as the 1^1S_0 and 1^3S_1 states, the number of photons that must be emitted is an important factor in determining the lifetime. In effect it takes longer to emit more photons. The 1^1S_0 state, which has C equal to +1, decays into two photons in about 10^{-10} second. The 1^3S_1 state has the same probability of electron-positron collision, but because C is −1 for this state it must decay into three photons; the lifetime is about 1,000 times longer.

T he positronium system can be described completely and with remarkable accuracy by the theory called quantum electrodynamics, or QED. The theory attributes the force between two electrically charged particles to the exchange of virtual photons emitted by one particle and absorbed by the other. The photons are said to be virtual because they can never be detected directly in the laboratory. The probability of an electron's emitting a virtual photon is proportional to α, the constant in Coulomb's law, which again has a numerical value of about 1/137.

To calculate the force between an electron and a positron one can begin by determining the force that results from the exchange of a single photon; the result is proportional to α. For more precision one can include the possibility of two photons being exchanged; this calculation yields a small correction to the initial result, proportional to α^2. The possible exchange of three photons gives rise to a still smaller correction proportional to α^3. To attain an exact result would require making an infinite series of such calculations, but because α is quite small the corrections quickly become negligible; considerable accuracy can be achieved by including only the first few terms in the series.

In recent years a theory of the color force between quarks has been developed; it is patterned after QED and is called quantum chromodynamics, or QCD. Just as QED describes the interactions of electrically charged particles, so QCD describes the forces that arise between particles that bear a color charge. Whereas there is just one kind of electric charge, however, there are three kinds of color charge, usually called red, blue and green. (The names, of course, have nothing to do with colors in the everyday sense.) Furthermore, whereas there is just one carrier particle for the electromagnetic force (the photon), there are eight carriers of the color force; they are called gluons. Quarks and antiquarks have color charges, and the forces between them come about from an exchange of gluons.

Perhaps the most important difference between QCD and QED is that the gluons themselves carry a color charge, whereas the photon is electrically neutral. A quark can continually emit and then reabsorb gluons, which effectively spreads its color charge over a region of space. When two quarks or a quark and an antiquark approach each other closely, the extended regions of color charge begin to overlap. Because of the overlapping the force between the particles is not as great as it would be between two point charges separated by the same distance. As the particles approach each other the color force between them, which is inherently quite strong, becomes weaker and takes on the same form as the Coulomb force: its strength is inversely proportional to the square of the distance. This property of QCD,

called asymptotic freedom, was discovered by Politzer and David Gross and Frank Wilczek of Princeton University. It is what led Appelquist and Politzer to expect the charmonium system to form nonrelativistic bound states.

The force between a quark and an antiquark when they are far apart is not well understood, but there is reason to think the force becomes constant, or independent of distance. If it does, the energy needed to separate a quark from an antiquark would increase without limit as the particles moved apart. Such a force law could explain why particles with a color charge (that is, quarks and gluons) have not been seen in isolation.

One reason the long-range color force has not been characterized in detail is that QCD, unlike QED, is not a theory in which calculations are currently practical. The constant α in QED is replaced in QCD by α_s, which has a substantially larger value; moreover, the value increases with the distance between the particles. Hence the probability that a single gluon will be emitted is large, and the probability for two or three or more gluons is not small enough for it to be neglected. The impracticality of doing calculations in QCD makes the quarkonium system all the more important for studies of the color force. QCD can serve as a guide in constructing models of the force; the models can then be tested experimentally on the quarkonium system. The results may in turn reflect on the validity and properties of the theory.

In order to study quarkonium it is necessary to create it. It seems the most effective way to do this is through the annihilation of high-energy electrons and positrons. The process is conceptually identical with the decay of positronium, but because of the higher energy it can proceed through a somewhat different mechanism. The electron and the positron annihilate to produce a single photon, an event forbidden in the decay of positronium because it cannot conserve both energy and momentum. These quantities must still be conserved in the high-energy annihilation, but the uncertainty principle of Werner Heisenberg in effect allows a momentary violation of energy conservation. Before the photon has existed long enough for its presence to be registered it decays into two or more new particles. It is a virtual photon.

In the final state created from the decay of the virtual photon all conservation laws must be obeyed. This condition can be met in a simple way:

the virtual photon gives rise to a particle and its corresponding antiparticle. In some cases the particle-antiparticle pair is merely another electron and positron, which can then move apart and be detected. The products of the collision can also be a quark and an antiquark, however, which do not escape unencumbered. Instead additional quark-antiquark pairs materialize and become bound to the original pair, so that what is ultimately observed is in all cases a set of hadrons.

The high-energy annihilations take place in the device known as a storage ring, where beams of electrons and positrons circulate in opposite directions within a toroidal vacuum chamber. Most of the work with charmonium we shall describe was done with the storage ring called SPEAR, which is a facility of SLAC. There are two interaction regions at SPEAR, and so two detectors can be operated simul-

taneously. The two we have employed are called Mark II and the Crystal Ball (see Figures 9.3 and 9.4). They provide largely complementary information.

The Mark II detector was built by workers from SLAC and the Lawrence Berkeley Laboratory of the University of California. It excels in measuring the energy of electrically charged particles, which are registered by a device called a drift chamber. In the chamber fine parallel wires extend through a gas-filled cylindrical volume. An electric potential of a few thousand volts is applied to adjacent wires. When a charged particle passes through the chamber, it ionizes the gas, and the liberated electrons then drift toward the nearest positively charged wire; the resulting current is detected electronically. By timing the current pulses the trajectory of the particle as it passes through 16 concen-

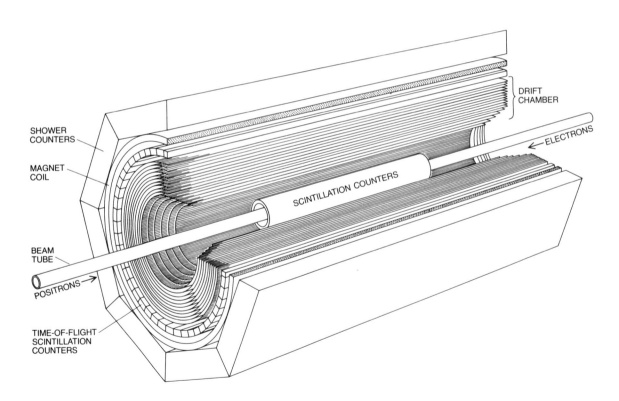

Figure 9.3 MARK II DETECTOR at the SPEAR storage ring is most effective in measuring the momentum of electrically charged particles. The main volume is a gas-filled chamber in which some 12,800 electrically charged wires are arranged in 16 concentric cylinders. A charged particle passing through the chamber ionizes atoms of the gas; liberated electrons drift to a positively charged wire, giving rises to a current that can be measured electronically. The entire chamber is permeated by a magnetic field that causes a charged particle to follow a curved path; from the radius of curvature the momentum of the particle can be deduced. Devices such as scintillation counters and shower counters aid in the identification of charged particles; they also detect photons, but with lower energy resolution than the Crystal Ball.

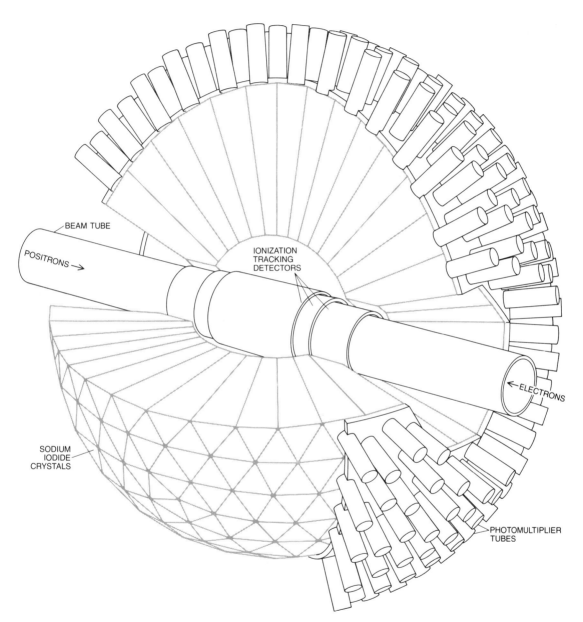

Figure 9.4 CRYSTAL BALL DETECTOR is mounted at a particle-storage ring; beams of high-energy electrons and positrons (or antielectrons) collide at the center of the detector and give rise to new particles, including quarkonium. The quarkonium then decays to yield still further particles, whose passage is recorded. The main elements of the detector are 732 crystals of sodium iodide. A crystal that absorbs a photon gives off light that can be registered by a photomultiplier tube. The direction of the photon and the amount of energy it deposits in the crystals provide information on the state of quarkonium that emitted the photon. Ionization tracking detectors aid in characterizing charged particles. The Crystal Ball was operated for about three years at SPEAR, a storage ring at the Stanford Linear Accelerator Center (SLAC). Later it moved to DORIS, a high-energy ring at the Deutsches Elektronen-Synchrotron (DESY) in Hamburg.

tric cylinders of wires can be determined to within about .2 millimeter. The entire drift chamber is immersed in a strong magnetic field, which causes a charged particle to follow a curved path. From the radius of curvature the momentum of the particle can be deduced.

The Mark II also has devices sensitive to high-energy photons, but these are detected with much greater precision by the Crystal Ball. Since the photon is neutral, it is not deflected by a magnetic field, and the Crystal Ball has no field. Instead the photon is absorbed in a dense crystal of sodium iodide, where it gives rise to a cascade of electron-positron pairs. The cascade causes the crystal to scintillate, or give off many photons at visible wavelengths, which are detected by a photomultiplier tube.

The Crystal Ball was designed and built by workers from Cal Tech, Harvard, Princeton and Stanford. The main component of the detector is an array of 732 crystals, which are arranged symmetrically around the interaction zone. The energy of a photon can be determined to within 2 or 3 percent, and the direction to within one or two degrees.

The discovery of the ψ meson at SLAC was made with the Mark I detector, the predecessor of the Mark II. At the time the detector was being employed to measure the rate of hadron production as a function of the energy of the incident electron and positron. If the quarks generated by the virtual photon initially act as free particles, the rate of hadron production should be proportional to the sum of the squares of the electric charges of all the quarks that can be created at a given energy. Over much of the energy range explored this assumption seemed to be confirmed by the measurements. When the colliding beams were set to an energy of precisely 3,095 MeV, however, the number of hadrons emitted abruptly increased a hundredfold. It was soon apparent that this sharp enhancement signaled the existence of a new kind of quark (see Figure 9.5). The exceptional height and narrowness of the peak indicated something else as well. At 3,095 MeV the assumption that the quark and the antiquark act as free particles fails badly. The charmed quark and antiquark that make up the ψ are not born free; they are tightly bound from the moment of their creation.

The ψ meson was not the first example of a sharp peak in the hadron production rate. Three other short-lived mesons had been observed in a similar way at lower energies: the ρ (rho) at 776 MeV, the ω

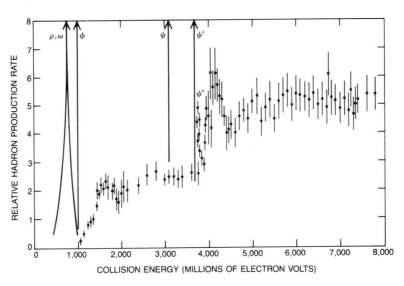

Figure 9.5 SHARP ENHANCEMENTS in the emission of hadrons (the generic term for the particles made up of quarks) signal the existence of quarkonium systems. The data for energies above 2,400 MeV were recorded at SPEAR; the lower-energy data come from storage rings in France, Italy and the U.S.S.R. The peaks at the left marked ρ (rho), ω (omega) and ϕ represent quasi-bound state of the lighter quarks. The ψ and ψ' are the 1^3S_1 and 2^3S_1 states of charmonium. The enhancement near 3,770 MeV is the 3^3D_1, state of charmonium, a quasi-bound state designated ψ''. The increase at about 4,000 MeV is due to the creation of charmed D mesons.

(omega) at 782 MeV and the ϕ (phi) at 1,020 MeV. One might suppose these mesons are bound states of the up, down and strange quarks, although they would presumably be relativistic bound states. This explanation is only partly correct: the ρ, the ω and the ϕ are considered quasi-bound rather than fully bound states. The distinction lies in whether or not the quarks in the hadron can eventually appear in its decay products. Consider the ϕ meson, composed of an $s\bar{s}$ pair. It can decay by mutual annihilation, but that is not the usual course of events (see Figure 9.6). Instead the s quark and the $\bar{s}$ antiquark merely move apart and a new u quark and $\bar{u}$ antiquark materialize, forming two new mesons with the quark composition $\bar{s}u$ and $s\bar{u}$. These mesons are

designated the K^+ and the K^-, and their total mass is 32 MeV less than the mass of the ϕ. The excess 32 MeV is converted into the kinetic energy of the K mesons.

The ψ might decay in a similar way except for an accident of nature. The meson composed of a c quark and a $\bar{u}$ antiquark is the D^0; its antiparticle, with the quark constituents $\bar{c}u$, is the $\bar{D}^0$. The mass of the neutral D mesons is 1,864 MeV, and so the mass of the particle-antiparticle pair is 3,728 MeV, which is 633 MeV greater than the mass of the ψ. Thus the ψ cannot decay into two D mesons because the decay would violate the conservation of energy. The c and the $\bar{c}$ cannot escape from each other but can decay only by annihilating each other to form

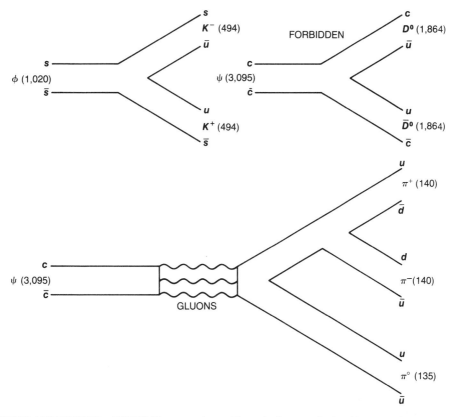

Figure 9.6 QUARK-ANTIQUARK SYSTEMS can be bound or quasi-bound according to their mode of decay. The ϕ (phi) meson is quasi-bound. It is the 1^3S_1 state of a strange quark s and a strange antiquark $\bar{s}$ and has a mass of 1,020 million electron volts (MeV). When the ϕ decays, the s and the $\bar{s}$ separate and a new quark-antiquark pair is formed; the products are the mesons designated K^+ and K^-.

The ψ (psi) meson is the 1^3S_1, state of a charm quark c and a charm antiquark $\bar{c}$. The lightest particles incorporating charm quarks are the D^0 and $\bar{D}^0$ mesons with a combined mass larger than the ψ. The ψ decays by annihilation into three gluons, which give rise to lighter quark-antiquark pairs and ultimately to ordinary hadrons.

gluons, which are then transformed into quarks and finally into ordinary hadrons. This process is comparatively slow and makes the lifetime of the ψ several hundred times longer than it would be otherwise (see Figure 9.6).

In analogy with the lowest-lying state of positronium the ψ was assumed to have zero orbital angular momentum. Because it is created from a virtual photon, however, its total angular momentum must be the same as that of the photon, namely one unit. Hence the ψ is the 1^3S_1 state of charmonium, the state with n equal to 1, one unit of net spin and no orbital angular momentum (see Figure 9.7).

Immediately after the discovery of the ψ it was suggested that the 2^3S_1 state, with the same quantities of net spin and orbital angular momentum but with n equal to 2, might also be a bound state. A search for it was undertaken by increasing the energy of the storage ring in increments of 2 MeV. Two weeks later the particle named the ψ' was

discovered at an energy of 3,684 MeV; it is now understood to be the 2^3S_1 state of the $c\bar{c}$ system. If its energy were 44 MeV greater, it could decay into D mesons and would be only quasi-bound.

If the charmonium model is correct, there must be at least five additional bound states of charmonium that might be detected in the decays of the ψ and the ψ'. Two of these states are the singlet S states that differ from the ψ and the ψ' only in that they have zero net spin as well as zero orbital angular momentum. Their designations are 1^1S_0 and 2^1S_0, and they are called charmed η (eta) mesons. To be precise, they have been given the names η_c and η_c' in analogy with the η meson, which is a $^1S_0$ state of the light quarks. The η_c and the η_c' were expected to have masses about 100 MeV lower than those of the ψ and the ψ' respectively.

The other three expected states were the triplet states with one unit of orbital angular momentum: 2^3P_0, 2^3P_1 and 2^3P_2. They were given the name χ

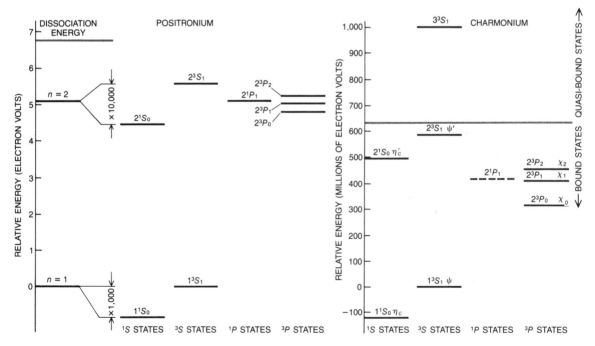

Figure 9.7 SPECTRUM OF ENERGY STATES is similar in positronium and charmonium, but the scale of the energy differences in charmonium is greater by a factor of roughly 100 million. The energy of a state is determined by the principal quantum number n and by the orientation of the particle spins and the orbital angular momentum. In positronium the various combinations of angular momentum cause only minuscule shifts in energy (shown by expanding the vertical scale), but in charmonium the shifts are much larger. All energies are given with reference to the 1^3S_1 state. At 6.8 electron volts positronium dissociates. At 633 MeV above the energy of the ψ charmonium becomes quasi-bound because it can decay into D^0 and $\bar{D}^0$ mesons. The 2^1P_1 state of charmonium has not been detected experimentally.

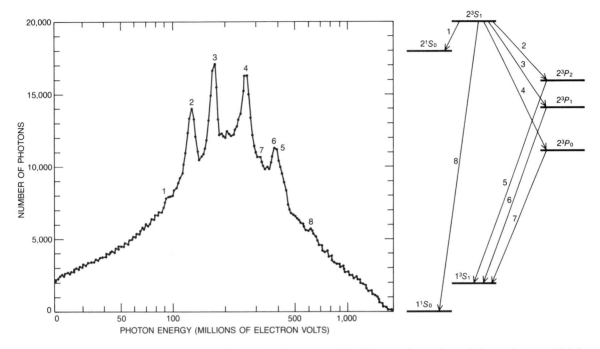

Figure 9.8 TRANSITIONS BETWEEN STATES of the charmonium system have been mapped with the Crystal Ball detector. Only the $^3S_1$ states can be made directly in electron-positron collisions; the others are formed when one of the $^3S_1$ states decays by emitting a photon, which is detected with the Crystal Ball. The peaks in the spectrum of photons are keyed by number to the transitions in the diagram at the right.

(chi) mesons by the experimenters who undertook to find them. They were expected to have an average mass about 200 MeV less than the mass of the ψ' and to be separated from one another by between 50 and 150 MeV. A sixth state is also predicted by the charmonium model; it is the singlet state with zero net spin and one unit of orbital angular momentum, namely 2^1P_1. A low expected rate of conversion between this state and the others, however, makes it unlikely to be detected, and indeed it has not yet been observed.

The singlet S states and the triplet P states of charmonium cannot be created directly in electron-positron annihilations for two reasons. First, with the exception of the 2^3P_1 configuration, they do not have the same total angular momentum as the virtual photon has, so that their direct production would violate the conservation of angular momentum. Second, for all five states the sum of the absolute magnitudes of the net spin and the orbital angular momentum is an even number (zero for the S states and 2 for the P states). For all the states C is therefore +1, in contrast with the virtual photon's

value of −1. For the ψ and the ψ', on the other hand, C is equal to −1, and so they can be formed directly. The $C = +1$ states appear only when the ψ or the ψ' decays into a photon and a lower-energy charmonium configuration, a process that obeys all conservation laws.

The first evidence of the χ mesons came from an experiment done at DORIS, a storage ring similar to SPEAR at the Deutsches Elektronen-Synchrotron (DESY) in Hamburg. In the years since then several other experiments based on a variety of detection methods have also seen the χ mesons. They have masses 135, 180 and 270 MeV less than the ψ' mass, in general agreement with the predictions of the charmonium model.

At SPEAR a group from the University of Maryland, the University of Pavia in Italy, Princeton, the University of California at San Diego and Stanford measured the masses of the χ states and the rate at which the ψ' decays to yield them. These workers made their measurements with an earlier detector based on sodium iodide crystals, detecting not the χ

particles themselves but the photon emitted during a transition from the ψ' to a χ. The energy of such a photon is sharply defined and is nearly equal to the difference in mass between the states. Measurements were also made with the Mark I, searching for events in which a χ decays into several charged particles. For example, a χ occasionally breaks down into two positive and two negative pions; from the energy and the angular distribution of the pions the mass of the χ state can be deduced. More recently higher-resolution measurements have been made by the Crystal Ball and the Mark II groups.

The angular-momentum quantum numbers of the χ mesons can be determined from the nature of the hadrons emitted in the decay of the particles or from the angular distribution of the photons. The early experiments showed that none of the three particles could be the $^1S_0$ states η_c or η_c'. Furthermore, if it was assumed that one meson had a total angular momentum of 2, another an angular momentum of 1 and the third angular momentum of zero, there was only one way to match the angular-momentum states with the observed masses. The most massive state had to be the one with an angular momentum of 2, the next state the one with an angular momentum of 1 and the least massive the one with an angular momentum of zero. This is the sequence of the 2^3P states of positronium. A more detailed study done with the Crystal Ball later confirmed the angular-momentum assignments.

The story of the η_c and the η_c' mesons is not as straightforward. Experimenters working at DORIS reported evidence for an η_c candidate in the early years of the exploration of the charmonium spectrum. They measured an anomalously large number of three-photon decays of the ψ meson in which two of the photons had an aggregate mass of 2,830 MeV; this mass is 265 MeV less than the mass of the ψ itself. They interpreted these events as the emission of a comparatively low-energy photon by the ψ, which was thereby transformed to the η_c state. The latter meson then gave up its mass of 2,830 MeV in forming the two additional observed photons. The signal seemed impressive, but there was a large background of extraneous events, and no other experiment was able to provide a convincing confirmation of the DORIS findings.

There was also a hint of an η_c' candidate based on four events recorded with the Mark I detector and an equal number from DORIS experiments. In these events the ψ' emitted a photon to become the η_c', which then emitted a second photon in the course of decaying to the ψ level. The mass of the η_c' candidate was 3,454 MeV, or 230 MeV less than the mass of the ψ'.

Neither candidate could be fitted easily into the charmonium model. The problem was that the masses were too low. In 1978 the expected mass of the η_c was calculated by Michael Shifman, Arkady Vainshtein, Michael Voloshin and Valentin Zakharov of the Institute of Theoretical and Experimental Physics in Moscow. From theoretical principles based on QCD they showed that the η_c must have a mass about 100 MeV less than the mass of the ψ, within an error of plus or minus 20 or 30 MeV. On this basis they concluded unequivocally that either the η_c candidate (with a mass 265 MeV less than that of the ψ) was not the true η_c meson or the simple charmonium model was not valid.

At this point the status of the charmonium hypothesis was uncertain. The three χ mesons had been found and their properties were in good agreement with the predictions of the model, but the candidate η_c states were at variance with the model in their mass and in certain other properties. It was not clear that the model was understood well enough for it to be trusted where it seemed to contradict the experimental evidence.

In 1978 the experimenters working with the Crystal Ball began collecting data, and by 1979 they had repeated the measurement of the decay of the ψ meson into three photons, the only decay mode in which the η_c candidate had been observed. The Crystal Ball was able to detect many more of these decays than had been recorded by the apparatus at DORIS and to resolve the energy of the photons with much greater precision (see Figure 9.8). The candidate was simply not there. The earlier experiment was almost certainly in error, although the reason is still not clear. It is possible there was a statistical fluctuation in the number of background events or a misunderstanding of the detector's response to them.

The η_c' candidate died a similar death. Experiments at DORIS and with both the Mark II and the Crystal Ball detectors at SPEAR searched the energy region where the candidate had been reported. Although all the new experiments were more sensitive than the earlier ones, they found no evidence of the η_c'. The particle with a mass of 3,454 MeV did not exist.

The state of uncertainty finally ended late in 1979 with the discovery by the Crystal Ball collaboration of the particle that appears to be the true η_c. The new η_c candidate has a mass of 2,980 MeV, or about 115 MeV less than the mass of the ψ, in accordance with the predictions of the charmonium model. The 2,980-MeV state was first observed as a small but significant peak in the spectrum of photons emitted in the decay of the ψ'. Later a peak with the same energy was noted in the photon spectrum from the decay of the ψ. The existence of the η_c particle was soon confirmed by the Mark II detector, which was able to record several of its decays into hadrons.

In August, 1981, the last of the five charmonium states with C equal to +1 was found close to where it had been predicted to be. The Crystal Ball collaboration reported the observation of another small peak in the spectrum of photons emitted by decaying ψ' mesons. The peak is at an energy of 3,592 MeV, some 92 MeV less than the mass of the ψ'; this is well within the range of masses predicted for the η_c'. With this discovery the η_c and η_c' puzzles seem to be solved, and the way is clear to apply the quarkonium model to the understanding of the color force.

Probably the best quarkonium system now available to experimenters is the $b\bar{b}$ system, or bottomonium. It should be superior to charmonium because it is appreciably heavier and is therefore a better approximation to a true nonrelativistic system.

When the Υ was discovered at Fermilab in 1977, it was interpreted as the 1^3S_1 state of bottomonium, directly analogous to the ψ, which is the same state of charmonium (see Figure 9.9). The discovery was accompanied by reports of the probable observation of the 2^3S_1 state, named the Υ', and the tentative observation of the 3^3S_1 state, the Υ'' state. The existence of three bound states of bottomonium (compared with two bound states of charmonium) was an important prediction of the model put forward by Eichten and Gottfried.

Because of the greater binding energy of bottomonium the array of associated bound states should be even richer than it is in charmonium (see Figure 9.10). For each of the Υ particles there should be a singlet S state of somewhat lower mass: they are the 1^1S_0, 2^1S_0 and 3^1S_0 states, which have been labeled

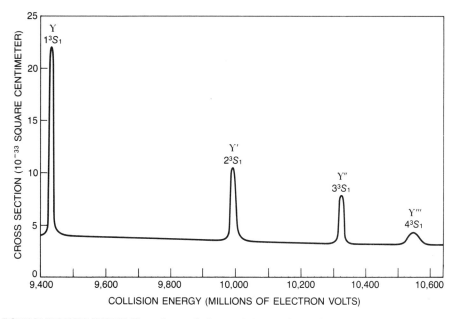

Figure 9.9 BOTTOMONIUM SYSTEMS, made up of a bottom quark and a bottom antiquark, are heavier than the corresponding states of charmonium and may reveal more clearly the nature of the interquark forces. Whereas two of the $^3S_1$ states of charmonium (the ψ and the ψ') are fully bound systems, there are three analogous bound states of bottomonium. They are designated Υ (upsilon), Υ' and Υ''. A fourth bottomonium system, the Υ''', is quasi-bound. The graph is based on data from the storage ring called CESR at Cornell University. The height of the peaks is given in terms of a cross section, which measures the probability of creating a particle.

η_b, η_b' and η_b''. Similarly, three χ_b particles should exist, representing the triplet of 2^3P states, that is, 2^3P_0, 2^3P_1 and 2^3P_2. In the bottomonium system, however, the $3P$ states should also be bound, whereas they are not in charmonium. Furthermore, the lowest-energy D states, with two units of orbital angular momentum, also ought to exist as bound systems. They are the 3^3D_1, 3^3D_2 and 3^3D_3 configurations.

As yet none of the bottomonium systems other than the Υ particles has been observed. The experiments are more difficult than those with charmonium, largely because the $b\bar{b}$ systems are formed at a much lower rate than the $c\bar{c}$ systems. The search is now under way in earnest at several laboratories. An electron-positron storage ring called CESR at Cornell has a nearly optimum energy range for creating bottomonium systems, and new storage rings capable of still higher energy have been built at both SLAC and DESY. In this connection the Crystal Ball has been removed from SPEAR and is being installed at an interaction region of the higher-energy ring at DESY. The coming generation of experiments should be able to observe the χ_b states and perhaps even the η_b.

W hat has been learned from quarkonium about the nature of the interquark force? Because the various states of quarkonium differ in the average separation between the quark and the antiquark, the energies of these systems convey information about the strength of the force over a range of distances. The experiments done so far constitute

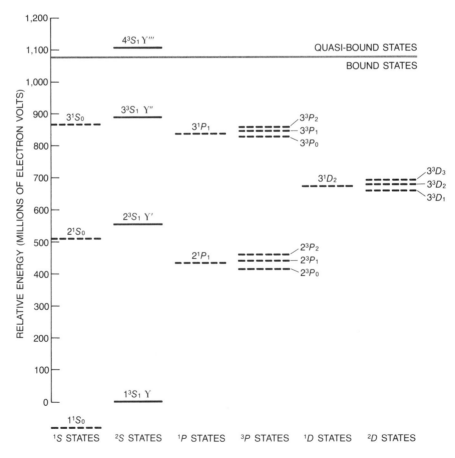

Figure 9.10 ENERGY LEVELS OF BOTTOMONIUM are expected to include more bound states than those of charmonium. In addition to the S and P bound states, certain bound D states, with two units of orbital angular momentum, may be detectable. Only the $^3S_1$ states have been observed.

measurements of the color force at distances ranging from 10^{-13} centimeter, which is roughly the size of an ordinary hadron, down to about 2×10^{-14} centimeter, a distance a fifth as large. From the measurements it is possible to construct models of how the force varies as a function of interquark distance (see Figure 9.11).

QCD suggests that the color force may vary inversely as the square of the distance when the quarks are close together but may assume a constant strength for quarks that are far apart. A plausible way to approximate such a force is to assume that it is merely the sum of an inverse-square force and a constant force. This model force law can be expressed mathematically by the equation $F = a/r^2 + b$, where F is the force, r is the interquark distance and a and b are constants to be determined by experiment. At very small distances the a/r^2 term is large and makes the dominant contribution to the total force. At large distances, on the other hand, a/r^2 is negligible and the force is essentially equal to the value of b.

The model force law can be tested by attempting to fit its predictions to the data for quarkonium. If the resulting curve has the right form, one can then ask what values of a and b give the best fit. It turns out that the form of the curve is consistent with the data, and the value of b is about 16 tons. In other words, two quarks attract each other with a force of at least 16 tons regardless of how far apart they are. With this in mind it becomes easier to understand why a quark has never been extracted from a hadron. It should be emphasized that we have not proved by these arguments that a constant force between quarks exists; a proof is hardly possible when the data extend only to distances of about 10^{-13} centimeter. What we have shown is that the existing data are consistent with a constant long-range force and that if it exists, its magnitude is about 16 tons.

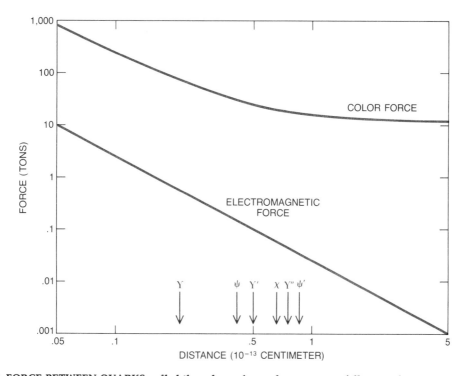

Figure 9.11 FORCE BETWEEN QUARKS, called the color force, seems to obey a law significantly different from the one that describes the electromagnetic force. The strength of the electromagnetic force between two particles varies inversely as the square of the distance between them; such a force law corresponds to a straight, sloping line on this logarithmic graph. At exceedingly short distances the color force also seems to follow an inverse-square law, but beyond about 10^{-13} centimeter the force may have a constant value independent of the distance. If the long-range force is constant, data from quarkonium suggest its magnitude is about 16 tons. The force law shown is based on a model devised by John Richardson.

The value of a in the equation for the force law is more difficult to specify. The reason is that a is not actually a constant but depends to some extent on r. The value of a is related to the probability that a quark will emit a gluon, the quantity we have designated α_g. Unlike α (the probability that an electron will emit a photon), α_s varies with distance, becoming smaller as the quark and the antiquark move closer together. For the range of distances explored in the quarkonium system α_s seems to have a value of approximately $1/5$, compared with $1/137$ for α.

Much of the interest in quarkonium stems from the close parallels between quarkonium and positronium and between QCD and QED. There is one phenomenon in systems governed by the color force, however, that has no equivalent in electromagnetic systems. It is a particle composed exclusively of gluons, bound together by their color charges. The analogous electromagnetic entity would be a bound state of two photons, but such a system cannot exist because the photon has no charge.

The hypothetical gluon bound states are generally called glueballs, but we shall call them gluonium systems to make plain their similarity to quarkonium. If they exist at all, the decay of heavy quarkon-ium is one of the processes most likely to create them. Just as the 1^3S_1 state of positronium decays into three photons, the 1^3S_1 state of charmonium (that is, the ψ particle) usually gives rise to three gluons. The gluons in turn create quark-antiquark pairs, which combine to form hadrons. About 10 percent of the time, however, charmonium decays into a photon and two gluons; it is in these events that gluonium might be most likely to form. The two gluons emitted in the decay have opposite values of color charge and could form a color-neutral bound system directly (see Figure 9.12).

It is not at all clear how to recognize gluonium if it is created. It is even possible that a particle might vacillate continually between quark-antiquark states and gluon states, in which case the concept of gluonium would not be very meaningful. If gluonium does have a fixed identity, one of its distinguishing characteristics might be the spectrum of angular-momentum states. Because the gluons have a spin of 1 rather than $1/2$ the gluonium state of lowest energy is expected to be an S state with a total angular momentum of either zero or 2. This is in contrast to quarkonium, where the S states have a total angular momentum of zero or 1. Another clue is that gluonium should be produced more abundantly in processes that are rich in gluons (such as

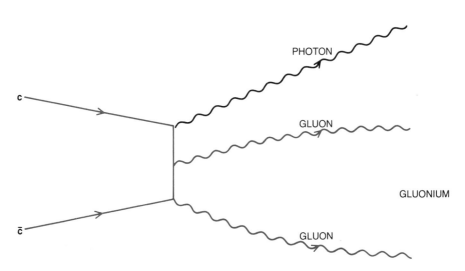

PHOTON

c

GLUON

GLUONIUM

c̄

GLUON

Figure 9.12 GLUONIUM is a hypothetical bound system made up entirely of gluons, the particles that ordinarily act as intermediaries transmitting the color force between quarks. Because gluons carry color charges the gluons are themselves subject to the color force and may form bound states. If gluonium exists, it is likely to be made in events that yield gluons in abundance. One plausible mechanism is the decay of a ψ meson into a photon and two gluons; under these circumstances the gluons would necessarily have opposite color charges and could become bound.

the decay of quarkonium into a photon and two gluons) that in other interactions of hadrons. Gluonium might also be made conspicuous by its failure to fit into any of the established families of mesons.

Daniel Scharre of SLAC discovered evidence in data from the Mark II detector of a curious mode of decay of the ψ particle. In roughly three decays in every 1,000 the ψ gave rise to a photon and a meson with a mass of about 1,420 MeV. Scharre identified the meson as the E, which is one of the 2^3P_1 states of the light quarks and therefore has a total angular momentum of one unit. The finding was surprising because the E meson is quite rare; although three per 1,000 is a small proportion of all ψ decays, it is considered a remarkably high rate of production for the E meson.

Michael S. Chanowitz of the Lawrence Berkeley Laboratory and others suggested that the solution might lie in a case of mistaken identity. The particle observed in the decay of the ψ had been identified on the basis of its mass and its most prominent mode of decay, which were the same as those of the E meson. Chanowitz pointed out that the new particle differed from the E in certain details of its production and decay. He suggested that what was being seen was not the E but a new particle, one with a total angular momentum of zero.

Chanowitz' arguments were proved to be correct when the Crystal Ball collaboration reported a measurement of the angular momentum of the 1,420-MeV particle; it was found to be zero. Since the particle could not be the E meson, it was given a new name: ι (iota) (see Figure 9.13). At the same time the Crystal Ball collaboration announced the discovery of yet another new particle among the decay products of the ψ. Like the ι, the second new particle appears in conjunction with a photon, but its mass is greater: 1,640 MeV. The second particle has been given the name θ (theta), and it has been found to have two units of total angular momentum.

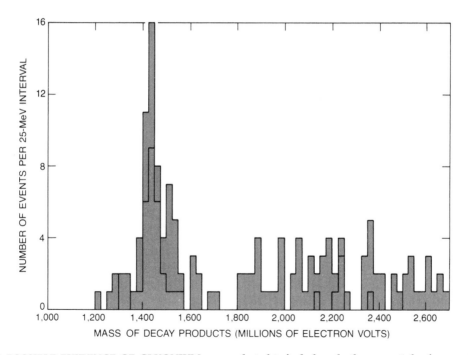

Figure 9.13 POSSIBLE EVIDENCE OF GLUONIUM may lie in the ι (iota) particle, which has a mass of 1,420 MeV and zero angular momentum. It does not readily fit into any of the well-established families of quark-antiquark mesons. It could be a bound state of two gluons, although other explanations are possible. Here decays of the ψ meson recorded with the Mark II detector have been se- lected to include only those events having a certain set of three particles among the decay products. The events were assigned to 25-MeV intervals according to the combined mass of the particles. The peak from 1,400 to 1,450 MeV is caused by the ι. Gray bars represent a subset of the data selected by more restrictive criteria.

What are the ι and the θ particles? The question is not easily answered. If they are quark-antiquark states, the ι must belong to the $^1S_0$ family of mesons and the θ to the $^3P_2$ family. They cannot be any of the $n = 1$ states of these families because those states are already fully occupied with known particles. They could be $n = 2$ states. They could be exotic "double" mesons, composed of two quarks and two antiquarks. Of course, they could also be gluonium states. Each of the possibilities entails a distinctive spectrum of accompanying new particles. If experiments show that the spectrum is the one characteristic of gluonium, the ι and the θ mesons will be recognized as the first examples of a new form of matter: particles embodying the interquark force in the absence of quarks.

POSTSCRIPT

Since we wrote this chapter, progress in the understanding of quarkonium has been remarkable. This is particularly true for the $b\bar{b}$ system, bottomonium, where many additional states have been observed by a number of experimental groups at Cornell and DESY. In addition, two gluonium candidates at 1440 and 1720 GeV have been studied extensively.

The best quarkonium system available to experiment is still bottomonium. It is superior to charmonium because it is three times heavier and is more nearly a true nonrelativistic system and therefore more easily calculable. To study these states, the SLAC Crystal Ball detector was moved to DESY in 1982 and installed in the newly commissioned DORIS e^+e^- storage ring, which had been optimized to run in the upsilon (1S) to upsilon (4S) energy regime.

The trip from SLAC to DESY was an interesting and exciting mix of land, sea and air travel. The main body of the detector, the sodium iodide crystal array, was transported by the U.S. Air Force in a giant C5-A cargo plane. The sensitivity of the sodium iodide and its large size made the flight technically very tricky. Because of the close cooperation between the Air Force and Crystal Ball experimenters, the flight went smoothly and the detector arrived unscathed.

With the four years of data-taking finished at DORIS, the Crystal Ball experiment essentially disbanded. The detector is still functional and available to any other collaboration interested and qualified to operate it.

Also, since 1982, both the 2^3P_j and the 3^3P_j states have been observed in the bottomonium system. These P states have masses and other characteristics very close to that predicted by theory, which is a great success of the quarkonium model. Most such models assert that the long-range force of QCD, the confining force, is the result of scalar exchange (rather than vector exchange) from the observed splitting of the bottomonium P states. That result is fundamental.

Understanding gluonium and establishing a set of criteria for its existence have also progressed. We are now almost certain that both the eta (1440) and f_2 (1720) are largely gluonium states, with some presently uncertain mixture of normal $q\bar{q}$ as well. Vacillation between the gluonic and quark states of gluonium appears to be occurring for these mesons. A theory describing them is now needed to understand the fundamental questions of their underlying QCD structure.

The study of toponium, the $t\bar{t}$ system of bound states, should provide the next quarkonium for investigating the color force. Here the heavier toponium samples the color force at much smaller distances than the psi and upsilon systems. It is in this small-distance region that parameterizations of the color force for psi and upsilon data differ from each other.

However the existence of the t quark and toponium has yet to be established. We know from experiments that the mass of toponium must exceed 46 GeV, and unconfirmed preliminary evidence suggests the mass of toponium might be between 60 and 100 GeV.

There are two new electron-positron colliding-beam facilities that might be able to study the toponium system. The first, the SLAC Linear Collider (SLC), is not a storage ring, since the electron and positron beams collide only once and are then discarded. The SLC is more efficient because the electron and positron beams can be made very small so that the probability of interaction is much higher than in a storage ring. We expect that in the future, very-high-energy electron-positron colliders will all be constructed on this principle. The SLC, which will produce toponium if its mass is less than 100 GeV, started operation in 1989.

A second electron-positron colliding beam facility, LEP, is nearing completion at the CERN laboratory in Geneva, Switzerland. It is a 27-km.-circumference storage ring over a hundred times larger than any similar existing machine. It is scheduled to operate in 1989 at 100 GeV, increasing to 200 GeV by the early 1990's.

Particles with Naked Beauty

The fifth quark, embodying the "flavor" known as beauty, has now been seen in combination with an antiquark not of its own flavor. The beauty of the new composite particles is accordingly exposed.

. . .

Nariman B. Mistry, Ronald A. Poling and Edward H. Thorndike
July, 1983

The ultimate building blocks of matter are currently thought to be the small set of indivisible particles called leptons and quarks. The world we normally experience is composed almost entirely of one kind of lepton, the electron, and two kinds of quark, arbitrarily labeled "up" and "down" (or simply *u* and *d*). The fractionally charged quarks, which are distinguished from one another by the quantum properties known as flavors, are bound together in different combinations to constitute the particles of the atomic nucleus: the proton (which has the composite structure represented by the letters *uud*) and the neutron (*udd*).

A complete description of all the known subatomic particles, including those observed only fleetingly in high-energy particle accelerators, requires more of these fundamental constituents. Six kinds of lepton are recognized, falling naturally into three doublets, or pairs. Until recently four kinds of quark were also well established, forming two similar doublets. In addition to the *u* and *d* quarks of ordinary matter a second pair of quarks is needed to embody the special flavors termed strangeness (*s*) and charm (*c*). This article summarizes the experimental evidence for a fifth kind of quark: the bottom, or *b*, quark, which embodies the flavor referred to as beauty.

The first indication of the existence of quarks endowed with beauty came six years ago as the result of experiments in which a beam of high-energy protons was directed against a stationary target. Among the products of the bombardment were a number of massive two-quark particles that appeared to consist of a new quark, the *b* quark, bound to its own antiparticle, the *b̄* quark. In such a particle the opposite flavors cancel and the particle is said to be flavorless; in other words, the observed particles were judged to have hidden, or concealed, beauty.

For the past few years our group at Cornell University has taken another approach to the search for particles with beauty. In our experiments oppositely directed beams of high-energy electrons and positrons are brought into collision; the colliding particles of matter and antimatter annihilate each other in a flash of electromagnetic radiation from which new particles can materialize (see Figure 10.1). With this method we have succeeded in creating particles in which the *b* quark is apparently bound to an antiquark of a different flavor. In these particles the beauty is no longer hidden; indeed, it is said to be naked. The observation of particles with naked beauty is expected to yield new information on the relations among the fundamental particles. It is also expected to focus renewed attention on the search for yet another quark to complete the third quark doublet. The predicted but so far unobserved sixth quark is labeled the top, or t, quark, and its flavor is referred to as truth. Before telling the story of the discovery of the beauty quark we shall review in somewhat greater detail the evolution of the lepton/quark description of matter.

The three lepton doublets consist of the electron (e^-) and its uncharged partner, the electron neutrino (v_e), the muon (μ^-) and its neutrino (v_μ) and the tau (τ^-) and its neutrino (v_τ) (see Figure 6.2). For each lepton there is an antilepton, and so the full membership of the class also includes the positron (e^+) and its antineutrino $(\bar{v}_e)$, the positive muon (μ^+) and its antineutrino $(\bar{v}_\mu)$ and the positive tau (τ^+) and its antineutrino $(\bar{v}_\tau)$. The leptons are very small, no more then 10^{-15} centimeter in diameter, and are probably pointlike. Although the charged leptons behave alike, their masses are quite different. The muon is 206 times as massive as the electron, and the tau is almost 3,500 times as massive as the electron. The neutrinos are very light and may actually be massless.

Each of the three quarks doublets (u and d, s and c, b and t) is made up of one quark with charge $2/3$ and another with charge $-1/3$. Like the leptons, the quarks seem very small and are usually thought of as point objects. Similarly, for each quark there is an antiquark, with opposite charge and opposite flavor. For example, since the b quark has charge $-1/3$ and -1 unit of beauty, its antiquark, the $\bar{b}$, has charge $+1/3$ and $+1$ unit of beauty.

With 24 building blocks (leptons, antileptons, quarks and antiquarks) it is quite remarkable that under most circumstances matter can be accounted for with only one lepton doublet (e^-, v_e) and one quark doublet (u, d). In addition to the three fundamental particles already mentioned as the main constituents of ordinary matter the fourth member of these first two doublets, the electron neutrino, is emitted by matter as a by-product of natural radioactivity.

Quarks are bound in other combinations to make up many particles besides protons and neutrons. Two possible combinations exist: groups of three quarks, called baryons, and quark-antiquark pairs, called mesons. Together the two classes of particles built from quarks are called hadrons. Hadrons, although they are small, are not pointlike. Their diameter of about 10^{-13} centimeter reflects their composite structure. Because of the strength of the force that binds quarks together, quarks are not detected in isolation. Physicists learn about quarks by studying hadrons. More than 100 different hadrons are known; most of them, however, are seen only in high-energy collisions produced by particle accelerators.

Of the four forces of nature three—the strong force, the electromagnetic force and the weak force—play important roles in the interactions of leptons and quarks. The fourth force, gravity, which is important in the macroscopic world, is negligible on the scale of quarks and leptons. In the modern view of the forces two particles interact by exchanging entities called gauge bosons. These interactions are illustrated schematically in Figure 10.2. In the figure tune goes from left to right. The exchanged objects behave somewhat like particles, but they do not have well-defined masses and they exist very briefly.

The strong force acts only among quarks; it is transmitted by the gauge bosons called gluons, which are so named because they are the "glue" that binds quarks together into hadrons. The electromagnetic force acts between any pair of charged particles; it is transmitted by the photon, which is the massless quantum of electromagnetic radiation. The weak force affects both quark and leptons; it is transmitted by the gauge bosons designated W^+, W^- and Z^0 (also known as intermediate vector bosons). Since the strong force is the strongest of the three forces, it plays the dominant role in quark and hadron processes, unless it is inhibited for some reason. In contrast, the weakness of the weak force, in the range of energies currently being examined, ensures that it is significant only when the strong and electromagnetic forces are suppressed. As the available energy increases, the weak force becomes stronger, until it eventually becomes comparable in strength to the other forces.

The gauge bosons associated with the respective forces act as the agents in all processes in which those forces are involved. The strong bonding of a d and a $\bar{u}$ quark to form a negative pion, or π^- meson, for example, is attributable to the exchange of gluons between the two quarks. Similarly, the electromagnetic force between two electrons is mediated by the exchange of photons. Another kind of electromagnetic interaction is observed when an

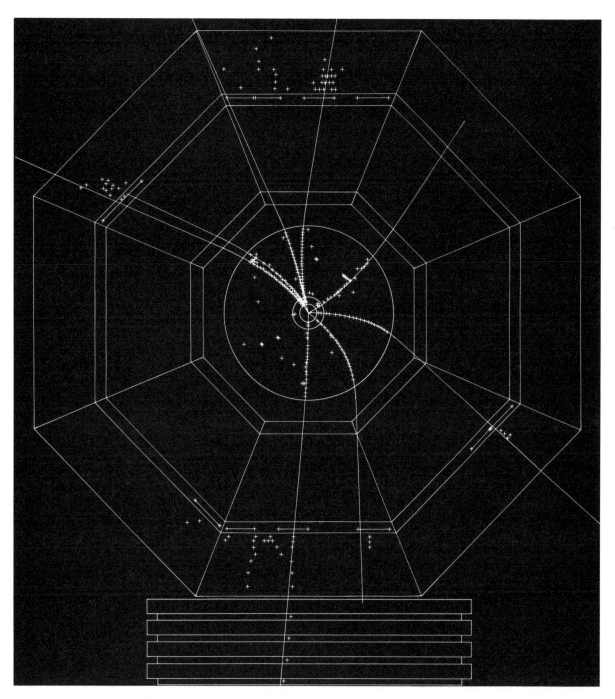

Figure 10.1 BEAUTY-FLAVORED MESONS produced the particle tracks represented in white in this computer-generated display. The head-on collision between an electron and a positron took place inside the CLEO detector. From the resulting annihilation there materialized a massive unstable meson called the upsilon (Υ‴), consisting of a beauty quark (b) loosely bound to its antiquark (b̄). In the subsequent decay of the Υ‴ two beauty-flavored mesons, B and B̄, were formed, and they decayed into an assortment of other particles. The passage of the electrically charged decay products through various parts of the detector was recorded as a pattern of "hits" (*white crosses*). (Display by Allen Beechel.)

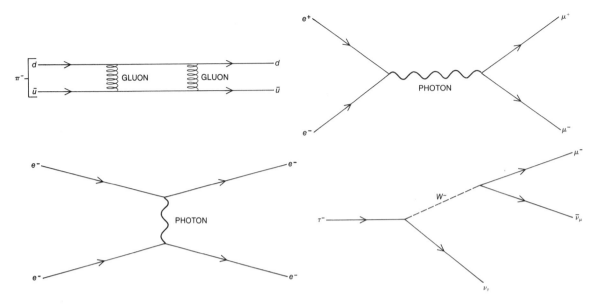

Figure 10.2 THREE FORCES in the interaction of sub-atomic particles operate by the exchange of gauge bosons. The gluon, the gauge boson of the strong force, binds the quarks together into hadrons; in this case the *d* and the $\bar{u}$ quark that constitute a negative pion, or π^- meson, are tightly bound by the exchange of a gluon (*upper left*). The photon serves as the carrier of the electromagnetic force; when two electrons approach each other, they interact by exchanging a photon (*lower left*). In another kind of elec-tromagnetic interaction an electron and a positron collide and annihilate each other to form a photon, which can then materialize into new particles, such as a muon-antimuon pair (*upper right*). When a negative tau lepton decays, it can transform into a tau neutrino by the emission of the intermediate vector boson W^-, one of three such carriers of the weak force, which decays into one of several possible combinations, such as a negative muon and a muon antineutrino (*lower right*).

electron and an positron collide and annihilate each other to form a photon, which promptly material-izes into a positive muon and a negative muon. The weak interaction is responsible for processes such as the decay of a tau lepton, in which the tau is trans-formed into a tau neutrino by emitting a W^- boson. The W^- in turn decays into a negative muon and a muon antineutrino.

The heavier quarks and leptons are unstable and decay into lighter quarks and leptons. The quark decays entail a change in flavor. Gluons and pho-tons are insensitive to flavor and cannot cause such changes; hence the quark and lepton decays are classified as weak interactions. In the decay of the *s* quark, for example, the *s* quark becomes a *u* quark when it emits a W^-. The W^- in turn can materialize into several possible pairs of particles, such as an electron and an electron antineutrino. One might also expect to observe the decay of an *s* quark into a *d* quark with the *s* quark emitting a Z^0 (the neutral counterpart of the W^+ and W^-), which could then decay into an electron and a positron. Careful searches have shown that such processes involving the Z^0 do not occur.

The need to incorporate this experimental fact into the theory led to the hypothesis that the *s* quark is the charge $-1/3$ member of a doublet with a charge $2/3$ quark of a new flavor, labeled charm. With two quark doublets the interplay of the quarks is such that the probability of the decay of an *s* quark into a nonstrange quark by the emission of a Z^0 vanishes. The existence of the *c* quark was pre-dicted on this basis almost 20 years ago, 10 years before the dramatic discovery of psi (ψ) particles, which are mesons composed of a *c* quark and a $\bar{c}$ antiquark [see "Electron-Positron Annihilation and the New Particles," by Sidney D. Drell; SCIENTIFIC AMERICAN, June, 1975]. This mechanism for sup-pressing certain strangeness-changing processes can be generalized to other flavors. In particular, if the *b* quark shares a doublet with a *t* quark, beauty-changing processes involving a Z^0 are forbidden. If there is no *t* quark, however, such processes must happen.

After the charmed quark was discovered in 1974 the lepton/quark description of matter seemed to be on a sound footing. The two lepton doublets and the two quark doublets recognized by that time accounted nicely for all the known particles. There was no compelling reason to expect any other leptons or quarks. The discovery of the tau lepton in 1975, however, suggested that there might be more quarks too (see Chapter 6, "Heavy Leptons," by Martin L. Perl and William T. Kirk). Evidence for a fifth quark was not long in coming. Indeed, there are many parallels between the discovery of the c quark and that of the b.

The first piece of evidence resulted from a 1977 experiment by a group of physicists at the Fermi National Accelerator Laboratory. The experimenters, led by Leon M. Lederman, then of Columbia University, bombarded a copper target with a beam of protons at an energy of 400 billion electron volts (GeV) and searched for muons produced in the collisions. When they observed a pair of oppositely charged muons, they considered the possibility that a new particle had been created and had subsequently decayed into the $\mu^+\mu^-$ pair. By measuring the momentum and the direction of each of the muons the experimenters were able to determine the mass of the new particle. The mass distribution that was obtained showed a broad peak at 10 GeV, indicating the presence of the new particle, which was named the upsilon (Υ). Lederman and his colleagues interpreted the shape of their peak as showing that they were in fact seeing two or more narrow peaks (probably three), smeared together by the limited experimental resolution (see Chapter 8, "The Upsilon Particle," by Leon M. Lederman).

The decay of the new particle into a $\mu^+\mu^-$ pair is an electromagnetic process. If the upsilon were to decay by the strong force acting at full strength, the electromagnetic decays would be insignificant in comparison and the $\mu^+\mu^-$ decay would not have been detected. Its observation suggested that the strong decay of the upsilon is somewhat inhibited. It was recognized that if the upsilon were a meson composed of a new quark and its antiquark ($b\bar{b}$), such a suppression could be readily explained. With enough energy the b and $\bar{b}$ quarks in the meson could separate, creating a new quark-antiquark pair and yielding "beauty-flavored" mesons, for example $b\bar{u}$ and $\bar{b}u$ (see Figure 10.3).

The latter decay is a strong-interaction process that would not be suppressed if it were energetically allowed. The observed suppression suggests that there is too little energy available for this process; in other words, the mass of the upsilon is less than the sum of the masses of two beauty-flavored mesons. The only alternative strong-decay mechanism is for the b and $\bar{b}$ quarks to annihilate each other and produce hadrons. This type of annihilation, although it is mediated by the strong interaction, is known to be suppressed. Indeed, it is because of the suppression that the electromagnetic decay is important. In the electromagnetic process the b and b quarks annihilate each other to form a photon, which then materializes into a $\mu^+\mu^-$ pair.

The suppression of the strong decay is expected to result in a lifetime for the upsilon that is considerably longer than the 10^{-23}-second lifetime of a typical strongly decaying hadron. Time intervals on this order are still too brief to be measured directly, but they can be determined from a relation between the lifetime of a particle and the precision to which its mass can be measured. If a particle lives forever, its mass can be determined with perfect accuracy. If its lifetime is very brief, measurements of its mass will give a distribution with a width related to the value of the lifetime; for a normal strongly decaying hadron the width of the mass distribution would be equivalent to an energy range of about .15 GeV. Lederman's experiment did not have sufficient resolution to measure the width of the upsilon's mass-distribution peaks; this determination awaited the arrival of a different technique.

The collision technique by which the upsilon family was discovered was not well suited to the detailed study of these particles. Upsilons are rarely created in proton-nucleus collisions, and when they are, they are usually accompanied by many other particles, making it difficult to follow the process. A far better technique for studying the upsilon family, and the b quark in general, is one in which beams of electrons and positrons are made to collide with each other. Sometimes an electron and a positron will annihilate each other to form a photon, which can materialize into a quark-antiquark pair, such as $b\bar{b}$. When the energy of the electron and that of the positron add up to the mass of an upsilon particle, there is an enhanced probability of the formation of a $b\bar{b}$ pair, which can bind to form an upsilon. Since upsilons decay primarily into hadrons, the experimental technique is to vary the energy of the colliding electron-positron beams and to measure the rate

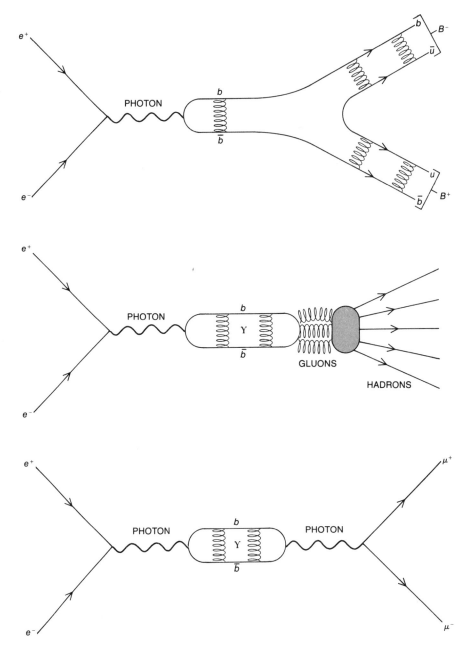

Figure 10.3 BEAUTY QUARKS ARE CREATED in electron-positron collisions when the photon resulting from the annihilation materializes into a $b\bar{b}$ pair. An additional quark-antiquark pair can be created and the $\bar{b}$ quarks can appear with the other quarks as beauty-flavored mesons (*top diagram*). At certain low energies the $b\bar{b}$ pair is created as a bound system: the upsilon meson. Then the quarks annihilate each other in one of two ways. The first type, actually inhibited, is by the strong interaction with the emission of gluons; the gluons in turn give rise to hadrons, which are observed in the aftermath of decay (*middle diagram*). In the electromagnetic decays of the upsilon the $b\bar{b}$ pair annihilates into a photon, which could form a muon-antimuon pair (*bottom diagram*).

of formation of hadrons. This technique was previously exploited at the electron-positron storage ring SPEAR, a facility of the Stanford Linear Accelerator Center (SLAC), in the discovery of the psi meson.

An experiment to search for the upsilons was carried out in 1978 with the electron-positron storage ring DORIS at the Deutsches Elektronen-Synchrotron (DESY) in Hamburg. Plots of the rate of hadron formation as a function of energy showed two sharp spikes at about 10 GeV, confirming the earlier indications that there were two or more distinct upsilon states. The DORIS machine had been strained to its limit in order to reach the energy of the second upsilon state, however, and it could go no higher. For this reason the third possible upsilon state was out of reach and, of greater importance, so were all $b\bar{b}$ systems in which the b and $\bar{b}$ were not tightly bound.

It was now crucial to see whether there is indeed a third bound state of $b\bar{b}$ and, at a slightly higher energy, another "resonance" where the b and $\bar{b}$ are only loosely bound and are produced with enough energy to separate and appear as beauty-flavored particles. Theorists had already predicted the energy levels of the bound $b\bar{b}$ system based on evidence obtained from the study of the $c\bar{c}$ system of the psi mesons. It was now up to the experimentalists to confirm or refute that picture and to study the behavior of the b quark through its decays and interactions.

At the time there was a growing sense of excitement among the physicists working on the preparation of the Cornell Electron Storage Ring (CESR). This machine, built under the direction of Boyce D. McDaniel and Maury Tigner, was designed in 1975 before the existence of the upsilon was known. It was a stroke of good fortune that the upsilon system lay right in the middle of the energy range planned for the CESR. Since the CESR was first turned on in April, 1979, it has been the site of most of the detailed work done on the b quark.

At the CESR a "bunch" of more than 10^{11} electrons and another bunch of roughly the same number of positrons travel in opposite directions in a ring with a circumference of 768 meters. The bunches circulate within a toroidal vacuum chamber maintained at a pressure of 10^{-8} torr (10^{11} times less than atmospheric pressure). Bending magnets and focusing magnets confine the particles to a circular path within the vacuum chamber and maintain the size and shape of the bunches. A constant supply of power to the beams compensates for the energy that is radiated away. Because of the comparatively low density of the electrons and the positrons in the bunches, there are very few head-on collisions at each crossing of the bunches. The high rate of crossings (400,000 per second) ensures, however, that there is an adequate rate of electron-positron annihilations. In the argot of particle physicists these collisions are "events." Hadronic events, those in which the collision products include hadrons, are observed a few times per minute.

The detailed study of the production and decay of b quarks requires the observation and study of hundreds of thousands of hadronic events. The rate at which these events happen is low, and so it is up to the experimenters to make sure their detectors are capable of recording every interesting collision. Furthermore, the detector must not miss many particles created in the events, because such particles may be crucial to understanding what happened in the production and decay of new particles.

A great deal of effort has been invested in the design and construction of systems for detecting and studying the products of such interactions. The challenges are formidable. Events may produce many particles, both charged and neutral, that can project in all directions. Furthermore, the rarity of interesting events makes high efficiency essential. The detectors must measure and record particles that pass through the detection zone in a few billionths of a second. At the CESR the electron and positron beams cross at two diametrically opposite intersection regions. The detectors surrounding the respective crossing points reflect two different approaches to the design of detection systems for such colliding-beam experiments (see Figure 10.4).

The CLEO detector, which occupies the south interaction region at the CESR, is a general-purpose detector built and operated by a group of 75 physicists (including the three of us) from seven institutions: Cornell University, Harvard University, Ohio State University, the University of Rochester, Rutgers University, Syracuse University and Vanderbilt University. It was designed to provide as much information as possible about each event. That includes detecting charged particles and measuring their momentum, measuring the energy and direction of photons and identifying specific kinds

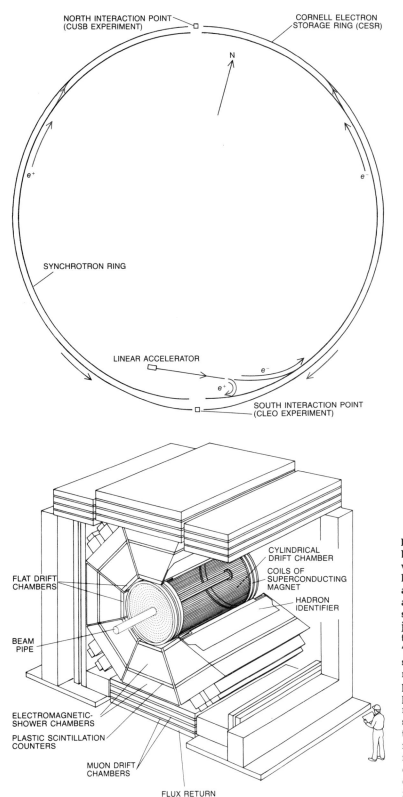

NORTH INTERACTION POINT
(CUSB EXPERIMENT)

CORNELL ELECTRON
STORAGE RING (CESR)

N

e⁺

e⁻

SYNCHROTRON RING

LINEAR ACCELERATOR

e⁻

e⁺

SOUTH INTERACTION POINT
(CLEO EXPERIMENT)

CYLINDRICAL
DRIFT CHAMBER

COILS OF
SUPERCONDUCTING
MAGNET

FLAT DRIFT
CHAMBERS

HADRON
IDENTIFIER

BEAM
PIPE

ELECTROMAGNETIC-
SHOWER CHAMBERS

PLASTIC SCINTILLATION
COUNTERS

MUON DRIFT
CHAMBERS

FLUX RETURN

**Figure 10.4 SITE OF THE DISCOV-
ERY of beauty-flavored mesons. The
view at the top shows the overall
layout of the CESR facility. Electrons
and positrons are accelerated first in
a linear accelerator and then in a
synchrotron ring before being injected into the storage ring. The particles are confined to separate
"bunches," which circulate in opposite directions and cross at two diametrically opposed interaction
points, where detectors study the
products of the electron-positron annihilations. The view at the bottom
shows the structure of the CLEO detector. A superconducting solenoid
magnet sets up a strong magnetic
field in the central cylindrical
chamber, where the momentum of
each charged particle is determined
from the curvature of its trajectory.**

of particles, such as pions, kaons (s-flavored mesons), protons, electrons and muons. Because particles can emerge in any direction, it was necessary to surround the interaction point as completely as possible with detector components. As a result CLEO is a very large device, occupying a space six meters long, eight meters wide and nine meters high, and weighing almost 1.5 million kilograms (1,500 metric tons). Careful computer analysis of the thousands of measurements done by individual components makes it possible to develop a clear understanding of what happens in specific annihilation events.

At the heart of CLEO is a superconducting solenoid magnet one meter in radius and three meters long. Inside the magnet is a cylindrical detector, called a drift chamber, that tracks charged particles as they emerge from the collision. As many as 17 separate measurements of the position of a charged particle are recorded, enabling the experimenters to trace the particle's trajectory. By measuring the curvature of the path in the magnetic field the momentum of the particle can be determined. The process of track detection and reconstruction is a crucial element for almost all work with CLEO.

Outside the coil of the solenoid magnet are components that serve to identify different kinds of charged particles. There is a set of plastic scintillation counters, which emit a brief flash of light when a charged particle passes through them. A precise measurement of the time of the flash, combined with accurate knowledge of the time at which the beams cross, makes it possible to calculate the time needed for a particle to travel from the collision to the counter. From this interval the speed of the particle can be determined. The measurement of the speed combined with the momentum measured with the central drift chamber leads to the determination of the particle's mass. On this basis it is possible to differentiate pions, kaons and protons.

Figure 10.1 illustrates a typical CLEO event. By fitting tracks to the hits an automatic pattern-recognizing computer program determined the direction and momentum of each particle and in some cases identified the type of particle. For example, the track at the six o'clock position in the figure was attributed to an extremely penetrating muon because of the string of hits registered in a special set of detection chambers arrayed around the outside of the detector. The track at 12 o'clock was identified as a probable electron on the basis of information re-

corded about the rate at which it deposited energy in an inner set of chambers, together with the telltale "shower" pattern it left in the electromagnetic-shower chamber at the top of the detector. A complete analysis of such collision events relies on data from all the detector's components, including some not shown in the figure.

In the original configuration of the CLEO detector the speed of a particle was also measured by a set of Cerenkov counters, which detected the light emitted when a particle passing through them exceeded the velocity of light in the gas with which the counters are filled. (The particle in this case acts somewhat like an airplane breaking the sound barrier.) The Cerenkov counters were useful for identifying electrons. They have now been replaced by gas-filled detectors that measure the rate of energy loss of each charged particle as it travels through the gas. The information can be combined with the measured momentum of the particle to yield the particle's velocity and hence its mass. In this way protons, pions, kaons and electrons can be independently identified.

Electrons have their identity confirmed in yet another part of the detector. When the electron passes through matter, it loses energy by producing what is called an electromagnetic shower. The process begins when a fast electron is deflected by a nucleus and emits an energetic photon. When the photon passes close to another nucleus, it gives rise to an electron-positron pair. The electron and the positron in turn emit photons, and the process continues with the production of a shower of secondary electrons, positrons and photons until all the energy is absorbed. The CLEO electromagnetic-shower detector is constructed of alternative layers of lead (to provide a medium for the production of showers) and charged-particle detectors (to sample the energy and extent of the shower as it develops). The shape and the energy of the showers associated with tracks coming from the collision point are analyzed to identify electrons. Showers that are not associated with any track are recognized as photons.

The detection of muons is based on their ability to penetrate large thicknesses of material. A hadron passing through matter is likely to interact strongly, and an electron will generate a shower; a muon, however, loses energy very slowly and can travel through much material without stopping. To exploit this property CLEO is surrounded by a steel absorber about a meter thick. Covering the outside of the steel are flat drift chambers that detect all charged particles emerging from the absorber. A charged particle that is observed to penetrate the steel is identified as a muon.

The CUSB detector in the north intersection region of the CESR was designed with a specific purpose in mind. It was built primarily as a high-resolution photon detector by a group of physicists from Columbia University, Louisiana State University, the Max Planck Institute for Physics and Astrophysics at Munich and the State University of New York at Stony Brook. Its main component is a set of shower detectors consisting of sodium iodide crystals. With these detectors it is possible to make precise measurements of the energy of photons. CUSB is particularly well suited to the measurement of photon energies in the range that is expected to be characteristic of transitions among the upsilon states. The CUSB photon detectors also serve to identify electron showers. Close to the collision point CUSB has tracking chambers to detect charged particles. Parts of the detector are surrounded by steel absorbers and also by counters for detecting muons.

T he study of b-quark physics at the CESR began in late 1979. The upsilon-particle resonances Y, Y' and Y" were quickly found at both CLEO and CUSB by measuring the rate of hadron production as a function of energy. The three resonances appeared as narrow peaks at the respective energies of 9.46, 10.02 and 10.35 GeV. This was the first time the Y" was clearly observed as a separate resonance, an event that was announced to the world of high-energy physics in a Christmas card. Since then the resonances have been carefully studied through the observation of several hundred thousand events. Some transitions among the energy states have been observed, and as rarer transitions are identified it should be possible to map out the complete spectroscopy of the upsilon system.

Since the upsilon resonances Y, Y' and Y" are hadrons with both beauty and antibeauty, the flavors cancel and the hadrons are flavorless. Accordingly these hadrons are said to have hidden beauty. When such a hadron decays, it does so by the mutual annihilation of the b and $\bar{b}$ quarks, and in that case the decays of the b quarks themselves cannot be examined. At electron-positron energies greater than the equivalent mass of the upsilon states $b\bar{b}$ pairs can be produced with enough energy for the two quarks not to be bound, allowing them to move away from each other. As they separate, the strong

force between the quarks causes a light quark-antiquark pair ($u\bar{u}$ or $d\bar{d}$) to be created. The quarks group together to form a B meson and a $\bar{B}$ meson ($b\bar{u}$ and $\bar{b}u$, or $b\bar{d}$ and $\bar{b}d$). In the B meson the beauty is no longer hidden. It is this particle that is said to display naked beauty (or bare bottom). In contrast to hidden beauty, naked beauty must decay by the weak interaction. Therefore by studying the beauty-flavored B mesons one can study the weak interactions of the b quark.

The upsilon resonances were originally interpreted as being three different energy states of the bound $b\bar{b}$ system. It seemed reasonable to expect that other resonances would exist above the threshold for decay into beauty-flavored hadrons. A bump-hunting expedition in the energy region above the Y" early in 1980 yielded the expected resonance, at 10.58 GeV (see Figure 9.9). Unlike the first three upsilon resonances, which were all extremely narrow, the fourth resonance, the Y''', was quite wide, indicating that it was decaying by the unsuppressed route into B and $\bar{B}$; in other words, it was a source of beauty-flavored particles.

To establish clearly that this interpretation was the correct one it was necessary to find evidence for particles carrying the beauty flavor in Y''' decays (see Figure 10.5). As we have seen, quarks of one flavor can decay into those of another flavor only through the weak interaction. One class of unmistakable weak decays consists of "semileptonic" decays in which a meson decays into a muon or an electron, together with a neutrino and one or more hadrons. CLEO was well equipped for identifying both muons and electrons, and the photon detectors of CUSB were also suited to the identification of electrons; accordingly both groups of investigators set about measuring the rate of single-lepton production in Y''' decays in order to confirm the presence of the weak decays of beauty-flavored particles.

B ecause of the difficulty of detecting leptons and the low rate at which Y''' events were accumulated, it was some time before a definitive conclusion could be drawn from the lepton yields. Through the first half of 1980, however, it gradually became clear that the rate of lepton production for events at the Y''' energy is significantly greater than it is for events at neighboring energies. Before accepting the conclusion that we were seeing beauty-flavored hadrons, however, we had to consider the possibility that the leptons are produced by some

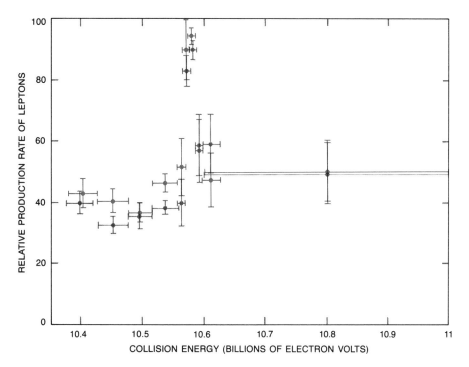

Figure 10.5 CLEARER INDICATION that beauty-flavored particles are produced in Y′″ decays is given by these plots of the rate of lepton production in electron-positron collisions as a function of the collision energy. The sharp increases in the production rate of electrons (*gray*) and muons (*color*) appear at an energy equivalent to the mass of the Y′″ resonance. The Y′″ evidently decays into some combination of products that in turn decay into a lepton and other particles. Such "semileptonic" decays are a telltale sign of a weak-interaction process and confirm that the Y′″ resonance decays to yield a $B\bar{B}$ pair. Symbols give range of experimental error.

mechanism other than the weak decay of B mesons. The observed distribution of lepton momenta is qualitatively what one would expect from the decay of a heavy meson such as the B and not from the decay of a lighter charmed meson. Since the leptons are produced singly and not in oppositely charged pairs, they cannot be the product of an electromagnetic process. When all alternative explanations were ruled out, the conclusion became clear: we were seeing naked beauty.

Although the lepton yields demonstrated the presence of naked beauty in Y′″ decays, we knew little about the B meson. Since the Y″ is below the threshold for $B\bar{B}$ decay and the Y′″ is above it, the mass of the B meson must lie between half of the mass of the Y″ and half of the mass of the Y′″, or in the range from 5.18 to 5.29 GeV. To demonstrate that the B meson is a real object and to measure its mass, it is necessary to reconstruct one from the debris of an Y′″ event. To do so calls for taking the tracks of an event in various combinations and determining whether any particular combination can be demonstrated to be the decay products of a B meson. The procedure is difficult because a B meson often decays into a large number of particles, some of which are neutral and can escape detection. Others emerge along the beam pipe and are not seen for that reason. What is more, each event has both a B and a $\bar{B}$ meson, and so the number of incorrect combinations can be extremely large.

By concentrating on decay modes of the B that involve only a few particles the CLEO group has recently succeeded in reconstructing B mesons (see Figure 10.6). To do so a sample of 150,000 events collected over a nine-month period was searched for the charmed particles known as neutral D mesons that decayed to form a kaon-pion pair. In some cases the D^0 was produced directly in the decay of a B meson. In others the B decayed into a charged excited state of the charmed meson, which in turn decayed into a D^0 and a charged pion. The reconstructed B's had all decayed into a D^0 or a charged excited state and one or two charged pions, giving a total of from three to five charged particles and no neutral ones. This is in marked contrast to the typical B, which decays into nine particles, three of which are neutral. The masses of the 18 B candidates cluster around 5.274 GeV.

Our mass measurement for the B demonstrates that the Y''', at 10.58 GeV, is only .03 GeV above the threshold for $B\overline{B}$ production (twice the B mass, or 10.55 GeV). When the Y''' resonance decays into $B\overline{B}$, there is just enough energy to make up the mass of the two mesons; as a result they are produced virtually at rest and with no accompanying particles to confuse the experimenter. Thus the Y''' is an ideal "factory" for making beauty-flavored particles: it is both an abundant source of B mesons and an uncluttered place to observe and study the decays of b quarks. There are many features of these decays we should like to explore. For instance, it is important to learn how quickly the b quark decays and what it decays into. If history is any guide, these experiments will lead to an improved understanding of the relations among the different flavors of quarks.

Although the b quark is expected to decay slowly on the time scale of elementary-particle processes, the decay seems very fast on a macroscopic time scale. The lifetime of the b is predicted to be between 10^{-13} and 10^{-14} second. In that brief interval a B meson traveling at a speed of 10 percent of the speed of light will move only .0002 centimeter. This distance is too small to be measured with the CLEO detector. So far the best that can be said from the CLEO data is that the lifetime is less than 10^{-10} second. A similar study done at the electron-positron storage ring named PETRA at DESY had the advantage of higher-energy (and therefore faster-moving) B's. In spite of an extremely low rate of B production, that experiment gave an upper limit on the B lifetime of 5×10^{-12} second.

What are the products of the decay of the b quark? The possibilities can be grouped into three classes. The standard view is that the b emits a W^- and is transformed into a c or a u. An alternative is for the b to emit a Z^0, transforming the b into an s or a d. This is a beauty-changing process involving a Z^0 and is forbidden if the b is part of a doublet with t. Finally, it is possible that the b decays in some entirely new and unexpected way, not involving the

emission of either a W^- or a Z^0. Let us consider each of these possibilities in turn.

In the standard view, with the b decaying into a W^- and a u or a c, the end products of b decay must include the particles into which the W^- itself decays. The options are $e^-\overline{\nu}_e$, $\mu^-\overline{\nu}_\mu$, $\tau^-\overline{\nu}_\tau$, $\overline{u}d$ and $\overline{c}s$ (see Figure 10.7). In general the quark options are three times likelier than the lepton options. By counting the possible outcomes and allowing for some necessary theoretical adjustments one can predict that the end products of b decay will include an electron about 13 percent of the time and a muon the same percentage of the time. The prediction agrees quite well with experimental measurements from CLEO and CUSB, lending support to the standard view of b decay.

This theoretical picture makes no prediction about whether the b quark decays more often to c or to u. Accordingly that is one of the most important things one can hope to learn about the b quark. The recent direct observations of D^0 mesons in B-meson decay demonstrates that the b quark does in fact decay to c at least some of the time. There are two procedures by which the decay preference of the b quark has been measured. The well-known dominance of the decay of c into s suggests the most obvious approach: one would expect B decays to produce more strange particles (kaons) if the b quark decays predominantly into c than if b decays into u. CLEO can detect both charged and neutral kaons. A careful study of kaon production shows there is a definite enhancement in the rate of kaon production in Y''' events compared with nonresonance events, which translates to a combined rate for charged and neutral kaon production of about 1.4 particles per B decay. Theoretically one would expect about .8 kaon for each B decay if b always decays into u and 1.6 if b always decays into c. The measurement suggests that the decay of b into c is dominant.

A better way to examine the question is to analyze the momentum distribution of the leptons from semileptonic B decays. Among the products of one of these decays there must be a lepton, a neutrino and one or more hadrons. If b decays into c, a charmed particle must be created. The lightest charmed particle is the D meson, with a mass of 1.87 GeV. Since this mass is quite large, it limits the amount of energy that can be carried away by the lepton as momentum. In contrast, if b decays into u, it is not necessary to create such a massive hadron, and it is possible for the lepton to have more mo-

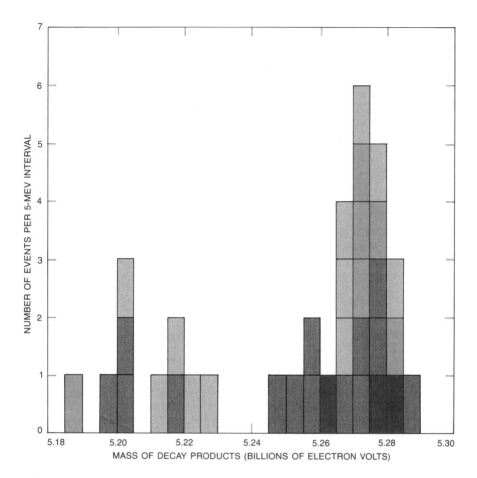

$B^{\pm} \to D^0 + \pi^{\pm}$

$B^0 \to D^0 + \pi^+ + \pi^-$

$B^0 \to D^{\pm} + \pi^{\mp}$

$B^{\pm} \to D^{\pm} + \pi^{\mp} + \pi^{\pm}$

Figure 10.6 RECONSTRUCTION OF BEAUTY-FLAVORED MESONS was accomplished by the CLEO group by examining the charged-particle tracks from the decay of the Υ''' resonance and tabulating them in various combinations according to four different models of the subsequent decay of a beauty-flavored meson (*key at left*). The resulting bar chart shows a peak in the vicinity of 5.274 billion electron volts. The specific decay modes considered are unusual in that the *B* meson decays into a small number of charged particles and no uncharged particles (other than the extremely short-lived D^0, or neutral charmed meson). The rarity of these decays and the difficulty of filtering them from other Υ''' decay processes is reflected in the fact that the small number of entries shown is the result of nine months of data collection, totaling 150,000 collision events.

mentum. By examining the production of high-momentum leptons and comparing these observations with what would be expected if *b* decays into *c* (or if *b* decays into *u*) it has been possible to demonstrate again that *b* decays primarily into *c*. In fact, it appears from this evidence that *b* decays into *u* no more than 5 percent of the time.

The standard view of the decay of the *b* quark, which has been the framework for the foregoing discussion, demands that the *b* quark share a doublet with the *t* quark. The discovery of the *t* quark is crucial to the confirmation of this model. The *t* quark has been sought in experiments at the electron-positron storage rings PEP (at SLAC) and

PETRA (at DESY). The failure to find the *t* may simply mean that the maximum energy of these searches (37 GeV) is too low to allow the production of bound $t\bar{t}$ states. The alternative is that the standard theory is incorrect and another theoretical description must be found.

If the standard theory is wrong and the *b* quark does not have a partner, one must consider the possibility that it decays by emitting a Z^0. (Remember, this process is not allowed in the standard theory.) Independent of whether the *b* turns into a *d* or an *s*, the decay products of the Z^0 must be present among the end products of *b* decay. The Z^0 can produce e^+e^-, $\mu^+\mu^-$, $\tau^+\tau^-$, $\nu\bar{\nu}$, $u\bar{u}$, $d\bar{d}$, $s\bar{s}$ or $c\bar{c}$. Here the e^+e^-, $\mu^+\mu^-$ options are very attractive because they yield clear signals. A search for indications of two routes of *b* decay, $b \to \mu^+ + \mu^- +$ hadrons and $b \to e^+ + e^- +$ hadrons, has yielded no evidence of such decays, leading to the conclusion that the Z^0 is not a major contributor to *b* decay.

Finally, there is the possibility that the *b* decays neither by W^- emission nor by Z^0 emission but in some entirely new way. Within the limits of rather general theoretical considerations only two additional kinds of decay are possible. The first is for the *b* to decay into a quark and two leptons from different doublets. The second is for the *b* to decay into a lepton and two antiquarks. If the *b* quark decays by the first of these exotic processes, with two leptons per decay, one would expect events with many muons, electrons, taus and neutrinos in various combinations. The taus in turn would decay, releasing part of their energy to neutrinos, which would be undetectable. If this kind of decay were significant, one would expect to observe frequent electron and muon production and considerable missing energy in *B*-decay events. By simultaneously measuring the yield of muons and electrons and the detectable energy in each Y''' event CLEO has shown that this kind of decay is not dominant.

In the second class of exotic *b* decays three more antiquarks than quarks are created (one that was originally bound to the *b* quark and two that were created in its decay). The presence of the three antiquarks results in the creation of one antibaryon per $\bar{B}$-meson decay. Similarly, the excess of quarks in $\bar{b}$ decay results in one baryon per $\bar{B}$ decay. Although there are many kinds of baryons (and antibaryons) that can be produced, all eventually decay into protons or neutrons (or into antiprotons or antineutrons). In CLEO protons and antiprotons can be identified, but neutrons and antineutrons escape undetected. If the *b* decays into a lepton and two antiquarks, one would expect *B*-decay events to yield, say, a muon, an electron, an antiproton or a large amount of missing energy. By simultaneously measuring the yields of muons, electrons and antiprotons, along with the detected energy per event, the CLEO experimenters have shown that this second kind of exotic decay is not dominant either.

In sum, all the evidence supports the view that the *b* quark decays by emitting a W^-. Furthermore, there are indications that most of the time it turns into a *c* quark rather than a *u* quark.

Although an understanding of the *b* quark and its interactions is developing, a number of open questions remain. What is the lifetime of the *b* quark? At present only upper limits have been established; they are far above the expected lifetime, however, and therefore provide no new information for evaluating the theory. What fraction of the time does *b* decay into *u* rather than into *c*? Data from CLEO and CUSB show that the fraction of *b* decays into *u* is small, but how small is it? Does the *t* quark exist? By showing that beauty-changing processes involving a Z^0 occur infrequently (if at all) and by showing that the unexpected processes that involve neither W^- nor Z^0 are also rare (if not absent) the CLEO experiments have ruled out almost all theories that can avoid having a *t* quark. Perhaps the *t* quark has not been found yet because it is too massive, and searches with higher-energy colliding-beam machines will find it. If the *t* does not exist, the theory of quark decays is in deep trouble.

Even if it is assumed that the *t* quark will ultimately be found, there remain some profound questions about the quark and lepton doublets. A perfectly satisfactory world can evidently be built of one lepton doublet (e^-, v_e) and one quark doublet (u, d). Yet at least three doublets of each type are thought to exist, and more may still be discovered. Why does nature behave in such an uneconomical way? How are the various doublets related and how do they differ? What does "flavor" really mean? Further study of the *b* quark may help to answer some of these questions.

POSTSCRIPT

When the electron-positron storage ring CESR started operating at Cornell University in 1979, it was clear that the trail through the tangled jungle of elementary particles pointed directly to the *b* quark.

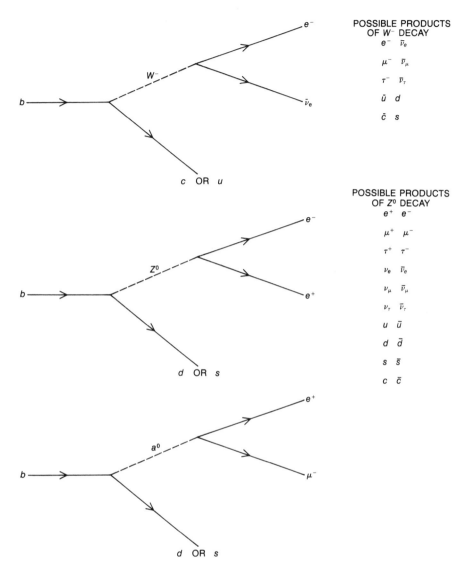

POSSIBLE PRODUCTS
OF W^- DECAY

e^- $\bar{\nu}_e$

μ^- $\bar{\nu}_\mu$

τ^- $\bar{\nu}_\tau$

$\bar{u}$ d

$\bar{c}$ s

POSSIBLE PRODUCTS
OF Z^0 DECAY

e^+ e^-

μ^+ μ^-

τ^+ τ^-

ν_e $\bar{\nu}_e$

ν_μ $\bar{\nu}_\mu$

ν_τ $\bar{\nu}_\tau$

u $\bar{u}$

d $\bar{d}$

s $\bar{s}$

c $\bar{c}$

Figure 10.7 THE POSSIBILITIES for the decay of the b quark are into a c quark or a u quark by emitting a W^- boson (*top diagram*). The W^- can then decay into five possible states, two of which ($e^-\bar{\nu}_e$ and $\mu^-\bar{\nu}_\mu$) are the semi-leptonic modes observed in the discovery of naked beauty. There is also a range of possible final states in the unlikely possibility a b quark decays into a d quark or an s quark by emitting a Z^0 (*middle diagram*). The decay of the b quark into a d quark or an s quark by the emission of a^0, an exotic neutral boson (*bottom diagram*), need not obey the rules that apply to conventional gauge bosons and so can decay into a negative muon and a positron.

The newly discovered upsilon resonances, with masses around 10 GeV, were believed to be bound states of the b quark and its antiquark $\bar{b}$. This was precisely the energy region for which CESR had been designed so the b quark became the focus of the new facility.

We described the early studies of the b quark and how it fit into the standard model of quarks and leptons. It is truly remarkable that the primary features of this twenty-year-old model have been seen in a succession of experiments at higher and higher energies. Its evolution can be followed through the chapters in this book. Sheldon Lee Glashow in 1975 reviewed the ideas demanding the existence of four quarks: the u, d, s and the new "charm" c quark. The discoveries of charmonium in 1974 and then charm mesons and baryons confirmed the deep insight of the physicists who predicted the c quark in the mid-1960's and foreshadowed the path that would lead to the b. Glashow hinted at the existence of a third family of quarks and leptons. The lepton member of this new family, the tau, appeared right on cue in late 1975 and is described in Chapter 6 by Martin Perl and William Kirk. The first hints of the existence of bottomonium states, bound states of the corresponding new "bottom" quark family, were announced in 1977 and are described in Chapter 8 by Leon Lederman.

The next logical step was to verify the new quark and study its properties. The task, which could best be performed with electron-positron colliding beams, fell to CESR, and to its two experiments, CLEO and CUSB. In spite of other higher energy rings, CESR was ideally designed to study the production and decay of the upsilon and its constituent b quark. At higher energies, large numbers of the new heavy particles containing b quarks are produced, but they are accompanied by additional particles because of the excess energy available. In contrast, for CESR running just above the threshold for production of the free b quark-antiquark pair, around 10.6 GeV, no additional particles are produced because there is little excess energy. The b-carrying mesons emerge alone with "naked" beauty.

We posed some open questions that remained in 1983, some of which have now been answered. For others, a few hints have appeared. The measured lifetime for decay of the b quark is about a picosecond (10^{-12} second). This is somewhat longer than expected but not so much longer as to require the rethinking of the overall picture. Our CLEO group has continued to identify and reconstruct the decays of B mesons; for example B mesons have been observed to decay into psi mesons, the so-called charmonium states. No evidence for the decay of the b quark into the u has been found, and at worst could account for 4 percent of all b decays. Every attempt to find discrepancies with the standard model has failed. These tests of the standard model, and the properties of the b quark in general, are still of great interest. A second storage ring, DORIS II in Hamburg, has joined Cornell in the study of b-quark decays. To extend our understanding beyond the limits of the present apparatus, a major upgrade of the CLEO detector is underway.

With the b quark fitting nicely into the framework of the standard model, the remaining missing piece of the third-generation puzzle is the "top" or t quark. It appears that the t is too massive to be produced at existing e^+e^- colliding-beam facilities; it must be heavier than 23 GeV. The search is escalating, with three new e^+e^- colliding facilities to be completed soon. The first, TRISTAN in Japan, is on line. It will have three detectors operating with beam energies initially just above those now available but ultimately to 70 GeV. In 1989 at the Stanford Linear Collider (SLC) and the Large Electron Positron Collider (LEP) at CERN, experimenters will attempt to produce t quarks in the decays of Z^0 bosons. This decay may be a prolific source but will be plagued by the confusion of extra particles and a wealth of different final states. Together, these experiments will cover energies up to 100 GeV. If evidence for the t is still not found, major revisions in our understanding of quarks and leptons may be necessary.

While there has been substantial progress in our understanding of quarks and leptons, the deepest questions we posed in 1983 are as troubling now as they were then: Why are there several varieties of quarks and leptons, and are they truly fundamental? The answers to these questions are at the heart of the quest for better experiments, higher energies and a deeper theoretical understanding of the sometimes "beautiful" world of particle physics.

SECTION

NOW AND BEYOND

. . .

Superstrings

If all elementary particles are treated as strings, a consistent quantum theory emerges that accounts for all four fundamental forces. The theory may transform our ideas about space and time.

. . .

Michael B. Green
September, 1986

The central paradox of the contemporary physics of elementary particles is the apparent incompatibility of its two main theoretical foundations. The first foundation is Einstein's general theory of relativity, which relates the force of gravity to the structure of space and time. This view of gravity has led to models of phenomena on a cosmic scale and to an understanding of the evolution of the universe. The second theoretical foundation is quantum mechanics, which can account for the atomic and subatomic world. Quantum theories have been formulated for three of the four known forces of nature: the strong, weak and electromagnetic interactions. Until recently there seemed to be little hope that Einstein's theory of gravity—the fourth fundamental force—could be united with the percepts of quantum mechanics. The basic difficulty is that such a unification seems to call for a radically new formulation of the laws of physics at the smallest distance scales; in such a reformulation the idea that space and time are continuous sets of points would have to be abandoned. Without a quantum theory of gravity and the conceptual revisions such a theory implies, a comprehensive description of all the forces of nature could not be realized.

In the past several years elementary-particle physicists have become optimistic that the theoretical impasse might be resolved. The optimism is based on striking developments in a new kind of theory known as superstring theory. In superstring theory, as in any other string theory, elementary particles can be thought of as strings. String theories thereby differ from all familiar quantum-mechanical field theories, such as the quantum theory of electromagnetism, whose quanta, or constituent particles, are pointlike. Since a string has extension, it can vibrate much like an ordinary violin string. The harmonic, or normal, modes of vibration are determined by the tension of the string. In quantum mechanics waves and particles are dual aspects of the same phenomenon, and so each vibrational mode of a string corresponds to a particle. The vibrational frequency of the mode determines the energy of the particle and hence its mass. The familiar elementary particles are understood as different modes of a single string.

Superstring theory combines string theory with a mathematical structure called supersymmetry [see "Is Nature Supersymmetric?" by Howard E. Haber and Gordon L. Kane; SCIENTIFIC AMERICAN, June, 1986]. Not only does superstring theory avoid the problems previously encountered in combining gravity with quantum mechanics, but also, in the process, the theory makes it possible to consider all four fundamental forces as various aspects of a single underlying principle. Furthermore, the unification of the forces is accomplished in a way determined almost uniquely by the logical requirement that the theory be internally consistent. These developments have led to an extraordinary revitalization of the interplay between mathematics and physics. Many of the deepest discoveries in modern mathematics are contributing to the understanding of the theory; in return string theories raise new issues in mathematics.

According to superstring theories, the standard laws of physics are approximate versions of a much richer theory that takes account of structure at an inconceivably small distance scale. The strings postulated by the theory are about 10^{-35} meter long, or some 10^{20} times smaller than the diameter of the proton. The differences between superstring theories and more conventional theories at such minute scales are essential to the consistency and predictive power of the theory.

For example, if one disregards gravity, it is possible to construct a unified picture of the strong, weak and electromagnetic forces in an ordinary field theory having pointlike quanta. The unified picture is the outcome of some underlying symmetry built into the theory, but there are many possible underlying symmetries. No theoretical reason is known for preferring one such symmetry to another. In contrast, in superstring theories gravity cannot be excluded, and the kind of symmetry needed for its inclusion in the theory leads to a natural prediction about the underlying symmetry that unifies the other three forces.

Since new concepts of space and time have long been expected from a quantum theory of gravity, it is worth mentioning how superstring theory could change our ideas about the geometry of the universe. It is not correct, strictly speaking, to regard the strings as independent particles moving in some fixed background space. In Einstein's theory of gravity, which superstring theory must approximate, space and time are unified in a four-dimensional continuum called space-time. The influence of the gravitational force is determined by the so-called curvature of space-time, which is analogous to the curvature of a two-dimensional surface such as the surface of a sphere. A particle moves along a

geodesic, or shortest path, in the curved space-time; on the sphere the analogue to such a path is the great-circle route between two points. The particle exerts a reciprocal influence on space-time, causing gravitational waves that can disturb the very geodesics along which the particle is moving. The equations of general relativity determine not only the paths of particles but also the structure of the space-time in which they are moving.

In superstring theory gravity is defined in a world expanded to nine spatial dimensions and time, making 10 dimensions in all. Again motion proceeds along geodesics, but the geodesics are surfaces of minimum area in 10 dimensions. Evidently six of the 10 dimensions must be hidden from view, thereby leaving only the four familiar dimensions of space-time to be observed. The six extra dimensions must be curled up to form a structure so small that it cannot directly be seen. The idea of unobservably small dimensions can readily be understood by considering a simple, two-dimensional analogy. A hose is a two-dimensional surface that appears to be one-dimensional when it is observed at scales too coarse to resolve its thickness. In superstring theory it is likely that the size of the six curled-up dimensions is approximately the same as the length of the string. The world appears to have three spatial dimensions in the same sense that the string acts like a point particle.

The expansion of the idea of geometry is not limited to adding six spatial dimensions. In ordinary general relativity a gravitational field is defined at every point in space-time. The equivalence of waves and particles in quantum mechanics requires that a gravitational wave, or disturbance in a gravitational field, be identified with a particle; the particle is called the graviton. Similarly, in string theory there should be a field that depends on the configurations of a string; such a field is called a string field. The number of possible configurations of a string in space is vastly greater than the number of points in the space; a string field would therefore be related to a new kind of geometry in an enormous extension of the idea of space, defined by all the possible configurations of a string. A stringlike particle should then be thought of as a "wavelike" disturbance in this huge space, just as a graviton is a wave in ordinary space.

Superstring theory can be historically traced to a theory called the dual-resonance model, which was developed in the late 1960's to explain the observed features of hadrons, or particles subject to the strong force [see "Dual-Resonance Models of Elementary Particles," by John H. Schwarz; SCIENTIFIC AMERICAN, February, 1975]. Although at the time an enormously successful quantum field theory of electromagnetism had been constructed, many theorists were becoming disillusioned with the general approach taken by quantum field theory. No such theory seemed able to account for the behavior of strongly interacting hadrons having a large spin, or quantized angular momentum.

It was in this context that Gabriele Veneziano, now at CERN, the European laboratory for particle physics, simply guessed a formula, unrelated to the formulas of quantum field theory, that expressed many features of hadron interactions. Subsequently Yoichiro Nambu of the University of Chicago, the late T. Goto, Holger B. Nielsen of the Niels Bohr Institute in Copenhagen and Leonard Susskind of Stanford University showed that applying Veneziano's formula is equivalent to describing the hadrons as strings. The harmonics of the string vibrations were supposed to correspond to the observed hadrons. Roughly speaking, the strings served to bind together the quarks that make up the proton, the neutron and other hadrons.

The original dual-resonance model could account only for particles, such as the pi meson, whose spin is an integer in fundamental units. Such particles are called bosons, and in quantum mechanics they are sharply distinguished from the fermions, such as the electron and the proton, whose spin is one-half times an odd integer. In 1971 a variant of the original theory that included fermions was developed by Pierre M. Ramond of the University of Florida, André Neveu of the École Normale Supérieure in Paris and John H. Schwarz of the California Institute of Technology. The variant, known as the spinning-string theory, was the precursor of supersymmetric theories.

Unfortunately both early string theories turned out to have several features that were considered serious liabilities at the time. First, the quantum-mechanical behavior of the original string theory for bosons makes sense only if space-time has 26 dimensions! For the spinning-string theory and indeed for the current superstring theories the corresponding number of dimensions is 10. Furthermore, the theories were plagued by the fact that the states of lowest energy of the string must be tachyons, or particles that travel faster than light. A relativistic quantum theory with tachyons is inconsistent. Finally, the theories required the existence of massless spin-1 and spin-2 particles, which did not correspond to observed hadrons. Their properties were more like those of the photon, the graviton and the so-called weak gauge bosons that carry the weak force. Although Joël Scherk, an exceptional physicist who died at a tragically young age, and Schwarz suggested string theory might be reinterpreted as a theory of gravity and the other forces, the inconsistencies lurking in such a theory seemed overwhelming.

In the early 1970's there was a great resurgence of interest in quantum field theories based on point particles. As recently as 20 years ago the only successful quantum field theory was the quantum theory of electromagnetism I mentioned above, known as quantum electrodynamics, or QED. Yet only a few years later two more highly successful quantum field theories had been developed: the electroweak theory, which gives a unified description for both electromagnetism and the weak interaction, and quantum chromodynamics, or QCD, which describes how quarks bind together to form hadrons (see Figure 11.1). These theoretical achievements were confirmed by remarkable experimental discoveries.

In all such theories the role of symmetry is paramount. The idea of symmetry in the laws of physics is expressed by a set of transformations that form a mathematical structure called a group. For example, the physical laws governing the behavior of an apparatus do not depend on its orientation in space; the laws are said to be symmetric under rotations about any of the three independent spatial axes. All such rotations belong to the three-dimensional rotation group designated $O(3)$; since rotations are specified by continuous angles, the group is a continuous symmetry group. It turns out that for every continuous symmetry in physics there is a conserved quantity or charge. The continuous symmetry of rotations in space gives rise to the conservation of angular momentum.

Many symmetries in particle physics are not related to ordinary space; instead they can be thought of as symmetries related to some so-called internal space. For example, in Maxwell's electromagnetic theory the internal space is regarded as a circle. Physical phenomena are independent of rotations about the circle just as they are independent of rotations in space. The conserved quantity associated with the symmetry is the electric charge.

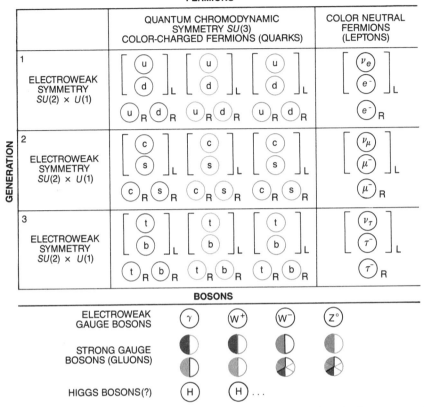

Figure 11.1 STANDARD MODEL of elementary particles combines quantum chromodynamics (QCD), which is the theory of the "color," or strong, force, and the electroweak theory. In superstring theory all these particles should arise as massless states of a string vibration. The fermions are grouped in the table into three "generations" of particles. The grouping displays the symmetry of the underlying theory much as the grouping of chemical elements in Mendeleev's periodic table displays the symmetry underlying atomic physics. Fermions include the quarks, which carry one of three color charges associated with the symmetry group $SU(3)$ of QCD, and the leptons, which carry no color charge. The color charges are represented here as red, green and blue. Quarks also carry electroweak "flavor," which is associated with the symmetry group $SU(2) \times U(1)$; six flavors are known and are indicated by the letters $u, d, s, c, b,$ and t. Leptons are subject only to the electroweak force. The leptons include the electron neutrino (ν_e), the electron (e^-), the muon neutrino (ν_μ), the muon (μ^-), the tau neutrino (ν_τ) and the tau (τ^-). Gauge bosons transmit the forces. The electroweak gauge bosons include the photon (γ) and the three massive particles W^+, W^- and Z^0. Eight strong gauge bosons, or gluons, carry color charges and anticolor charges, shown as solid color or as a color outline. There may also be Higgs bosons, which are responsible for the nonzero masses of particles. The subscripts L and R (for left and right) indicate handedness. For every particle except the neutrinos there is an antiparticle of the opposite handedness. The left-right asymmetry, most notable in the absence of right-handed neutrinos, is a signal that the weak interaction distinguishes among particles on the basis of handedness.

The richest and most interesting field theories are the ones in which there can be independent symmetry transformations at each point in space and time. Such symmetries are called gauge symmetries, and the theories are called gauge theories. The earliest and simplest example of a gauge theory is again Maxwell's theory of electromagnetism. A circle is associated with each point in space-time, and the laws of electromagnetism are not altered by independent rotations about each of the circles in this infinite set. This gauge symmetry is expressed by a symmetry group designated $U(1)$.

In any gauge theory there is a set of gauge particles that transmit the force between particles bearing a charge. In electrodynamics the gauge particle is a massless spin-1 particle called the photon. More general gauge theories with greater internal symmetry, such as the electroweak theory and QCD, also include massless spin-1 gauge particles. Such theories are known generically as Yang-Mills theories, after C. N. Yang of the State University of New York at Stony Brook and Robert L. Mills of Ohio State University. The forces described by Yang-Mills theories are nongravitational; they are called Yang-Mills forces. Finally, even Einstein's theory of gravity is a kind of gauge theory, but the spin of its gauge particle, the graviton, is 2.

One important aspect of a symmetry is that it can appear to be broken; the spontaneous appearance of a broken symmetry signals a change of state called a phase transition of a system of particles. For example, a lump of iron is not magnetized at high temperatures because the magnetic moments of the atoms point randomly in all directions. Since no direction is picked out, the system has rotational symmetry. When the iron is cooled, there is a change of phase: the atomic moments line up and the iron becomes magnetized in a particular direction. The rotational symmetry appears to be broken. Note that the rotational symmetry of the laws governing the forces on a microscopic scale remains intact; the appearance of a broken symmetry is characteristic of a system (the lump of iron) made up of a large number of particles.

Similarly, many theories of elementary particles, including superstring theories, require large gauge symmetry groups in order to give a unified account of diverse phenomena. Typically such large symmetry is apparent only at extremely high temperatures; the symmetry must appear to be broken at ordinary terrestrial temperatures if the theory is to be consistent with observations. For example, the electroweak theory describes a unified version of the electromagnetic and weak forces called the electroweak force, whose gauge symmetry is based on a group called $SU(2) \times U(1)$; the group is an extension of the group $U(1)$ associated with electromagnetism. At ordinary temperatures, however, one observes two forces, the electromagnetic and weak forces, which are entirely distinct. The symmetry associated with the unification of the two forces becomes apparent only at temperatures much higher than 10^{15} degrees Celsius.

The success of the quantum field theories of point particles gave a new lease on life to quantum field theory, and many physicists turned their attention to more ambitious schemes for unification. Such schemes were almost invariably based not on string theories but on more elaborate, so-called grand-unified symmetries built into quantum field theories with point particles. The grand-unified schemes, which ignore gravity, were associated with symmetry groups called $SU(5)$, $SO(10)$ or E_6. These large symmetries can break into smaller symmetries associated with the group $SU(3)$ of QCD and the group $SU(2) \times U(1)$ of the electroweak theory.

The temperatures at which grand-unified symmetries—and indeed the effects of quantum gravity—might become important are extraordinary: between 10^{30} and 10^{32} degrees C. According to current thinking about the origins of the universe, these temperatures were realized only between 10^{-43} and 10^{-38} second after the big bang. In spite of the brevity of this period, its implications for the subsequent evolution of the universe have been profound. In this way it turns out that the physics of the incomparably small is a key to the understanding of phenomena on a cosmic scale.

Many of the new grand-unified theories also incorporated supersymmetry, a symmetry that transforms bosons and fermions into one another and thereby unifies particles having integer and half-integer spin. In a supersymmetric theory there are equal numbers of bosons and fermions for any given mass. Several recent attempts have been made to combine Einstein's theory of gravity with supersymmetry. The resulting so-called supergravity theories belong to a new kind of gauge theory in which the gauge particle responsible for supersymmetry is called the gravitino; its spin is $3/2$. The most popular supergravity theory for a time was formulated in 11 space-time dimensions: the four ordinary space-time dimensions and seven additional spatial ones.

It is ironic from the current theoretical standpoint that one of the most important early discoveries in string theory was made in the process of studying higher-dimensional supergravity theories. In 1976 Ferdinando Gliozzi of the University of Turin, Scherk and David A. Olive of the Imperial College of Science and Technology in London suggested that the spinning-string theory could be made supersymmetric. The implications of this argument, however, were largely unrecognized, and work on string theory was virtually abandoned. It then lay dormant until 1980, when Schwarz and I (together with contributions from Lars Brink of the Chalmers Institute of Technology in Göteborg) began constructing and investigating the properties of string theories having space-time supersymmetry.

In order to understand why a solution to the problems of reconciling gravity with quantum theory has proved so elusive one must consider the implications of Heisenberg's uncertainty principle at distance scales of less than about 10^{-15} meter. According to that principle, the more precisely a spatial measurement is made, the less precisely the momentum or the energy of the system being measured can be known. The uncertainties of energy are realized as fluctuations at short distances; because energy and mass are equivalent, the energy fluctuations can be manifest as the creation of "virtual" particles (see Figure 11.2). Virtual particles and antiparticles can materialize out of the vacuum for a short time before they annihilate one another. The sea of virtual particles gives rise to multiparticle effects similar to the ones that arise for a substance such as a ferromagnet made up of many atoms.

When the spatial resolution is less than about 10^{-35} meter, the energy fluctuations become so huge that, according to general relativity, virtual black holes form. The energy of the fluctuations, about

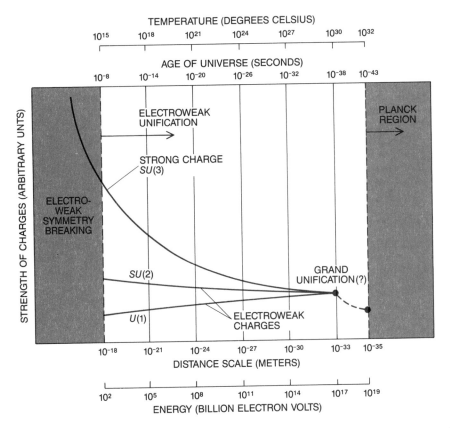

Figure 11.2 CHARGES associated with the electroweak force and the strong force depend on the separation between the particles carrying the charges. Virtual particles carry charges that can shield the separated particles, thereby causing the variation in the strength of the charges. If one extrapolates from the measured values of the charges at an energy of 100 GeV, the three charges appear to be equal in strength at about 10^{17} GeV. The extrapolation suggests a larger, "grand unified" symmetry prevails above that energy, in which the electroweak and strong forces become indistinguishable. The grand-unified energy, however, may not be distinct from the energy at the Planck scale (10^{19} GeV), in which case the force of gravity cannot be neglected.

10^{19} GeV (a GeV is one billion electron volts), is called the Planck energy, and the distance scale is called the Planck distance. Space-time must therefore be considered highly curved on small scales; in a sense it is foamy. This conclusion has disastrous consequences for the usual computational techniques of quantum field theory because it invalidates the notion of space-time as a smooth collection of points. In all such calculations one assumes the curvature of space-time is small; if the assumption does not hold, the calculations give nonsensical infinite results. This argument suggests that at short distances or, equivalently, at high energies either general relativity or quantum mechanics (or both) must be revised.

The corresponding argument cannot be made for strings, since superstring theory radically modifies the assumptions of general relativity at short distances. Indeed, as I outlined above, superstring theories may require a revision of general relativity at its most fundamental level: the idea of a curved space-time, which is central to Einstein's theory, may have to be extended to the infinitely richer space of string configurations. The current understanding of the theory, however, begins with a more primitive model, in which the strings are treated as independent particles moving in some fixed space-time background. Even with this restriction the quantum-mechanical treatment of a string leads to strong constraints on any superstring theory of gravity in higher dimensions.

There are two kinds of strings, open and closed. Open strings have endpoints. Conserved charges, such as the electric charge, that are associated with the Yang-Mills forces are tied to the endpoints. Particles associated with the vibrational states of an open string include the massless spin-1 gauge particles, but they do not include the graviton.

When open strings collide, they can interact by touching and joining at their endpoints to form a third string; the third string can then split apart to form two final strings. Similarly, the two endpoints of an open string can join to form a closed string. The vibrational states of a closed string include the massless spin-2 graviton. Thus in any theory with open strings there are also closed strings, and in any string theory with closed strings it is inconsistent to neglect the force of gravity. Accordingly, if the Yang-Mills forces, such as electromagnetism, are included in a string theory, they must be unified with gravity in an intimate way.

A kind of theory in which the Yang-Mills forces can be associated with closed strings was formulated by David J. Gross, Jeffrey A. Harvey, Emil J. Martinec and Ryan Rohm of Princeton University. Such a theory is known as heterotic, and it is the most promising kind of superstring theory developed so far. Its construction is quite strange. The charges of the Yang-Mills forces are included by smearing them out over the entire heterotic string. Waves can travel around any closed string in two directions, but on a heterotic closed string the waves traveling clockwise are waves of a 10-dimensional superstring theory; the waves traveling counterclockwise are waves of the original, 26-dimensional string theory. The extra 16 dimensions are interpreted as internal dimensions responsible for the symmetries of the Yang-Mills forces.

As a string moves it sweeps out a two-dimensional surface in space-time called a world sheet, just as a point particle sweeps out a world line (see Figures 11.3 and 11.4). In classical, or non-quantum-mechanical, general relativity a particle moves along the world line that minimizes the so-called action of the particle: its energy as it moves through time. The action is proportional to the length of the world line, and so a path of least action is a geodesic, or the shortest distance between two points in space-time.

The motion of a string is treated in an analogous way. In a non-quantum-mechanical approximation the string also moves in a way that minimizes its action. The action is proportional to the area swept out by the string, and so the world sheet must be a surface of minimum area. If time is regarded as a spatial dimension, the world sheet swept out by a closed string can be thought of as a kind of soap film that joins the string at its starting point and at the end of its path in space-time.

There is an enormous symmetry embodied in the condition that the motion of the string is determined by minimizing the area of its world sheet. The area is a geometric quantity that is independent of how the points on the two-dimensional sheet are labeled. No amount of distortion of the labeling can change the underlying geometry of the sheet, and physics is said to be symmetric under arbitrary relabelings of the world-sheet coordinates.

Because of this symmetry, there is no physical significance to distortions of the world sheet that lie in the two directions tangent to the surface of the sheet at any point. The only meaningful vibrations of the string are undulations of the world sheet

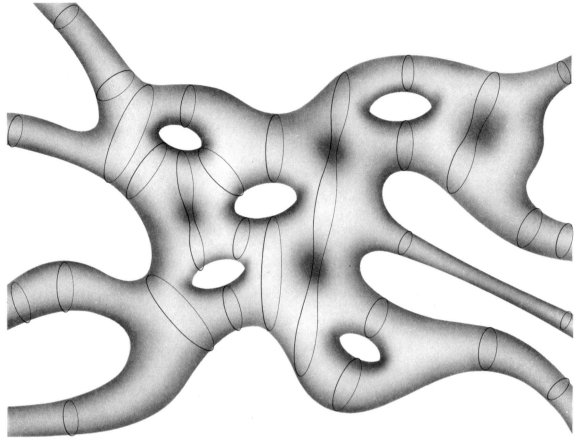

Figure 11.3 WORLD SHEET is swept out by stringlike particles as they move and interact in space-time. A string-like particle, like an ordinary string, has extension in only one dimension. In the diagram time is the horizontal axis; closed strings, or loops (*color*), enter from the left and leave on the right. The closed strings sweep out world sheets that are deformed cylinders. The diagrams for processes involving open strings are more complicated because the surfaces have boundaries traced out by the endpoints of the string. When two strings collide, they join to form a third string: two cylinders form a third cylinder. When strings split apart and rejoin, a hole is left in the world sheet. In quantum calculations all these splittings and joinings must be considered. The world sheet structure of these quantum mechanical interactions is like a doughnut with an arbitrary number of holes.

perpendicular to its surface. Hence if the string is moving in d dimensions, two directions of vibration have no physical reality. All real vibrations are transverse vibrations in $d - 2$ dimensions.

It is worth noting that such a constraint on the vibrational modes of photons, or waves in an electromagnetic field, is a hallmark of Maxwell's theory. There is no physical meaning to vibrations of the photon in the time direction, and longitudinal vibrations, which take place in the same direction as the wave is moving, are not possible for a wave traveling at the speed of light. The gauge symmetry of electromagnetism ensures the absence of such nonphysical vibrations.

From this point of view one might expect insurmountable problems in a string theory, since a string is made up of an infinite number of points. Each point along the string vibrates, and so the potential for nonphysical vibrations of a string is infinitely greater than it is for a point particle such

INITIAL STATE   FINAL STATE

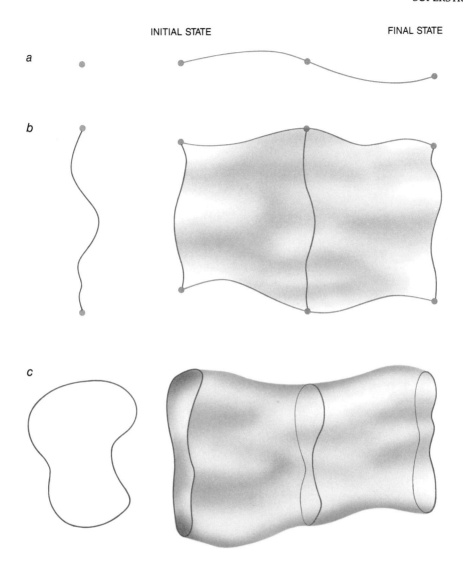

Figure 11.4 LEAST-ACTION PRINCIPLE determines the classical trajectory of a particle: if time is treated as a fourth spatial dimension, the trajectory is the line of minimum length in spacetime joining the initial state and the final one (*a*). Similarly, according to the least-action principle, a string sweeps out a world sheet of minimum area in spacetime. Cross sections are shown at the left. Open strings carry charges at their endpoints (*blue*) that define the boundaries of the sheet (*b*). Closed strings are loops, and so they have no endpoints (*c*). A string satisfying the least-action principle can vibrate in any combination of harmonic frequencies.

as the photon. The absence of such vibrations is guaranteed by the symmetry under coordinate relabelings. When the quantum mechanics of the string is considered, the same symmetry also introduces powerful constraints on possible string theories. It is in this sense that superstring theory is immensely elegant.

U p to now I have discussed the string as a classical particle. In quantum mechanics the motion of a particle is not precisely defined. When it moves through space, any particle tries, in effect, to take all possible paths between its initial state and its final state. The probability of each path is weighted so that a higher probability is assigned to paths of lower action; the classical path of least action is the most probable one. This formulation of the quantum behavior of a particle is called the method of summing over histories, and it was devised by Richard P. Feynman of Caltech. Its application to string theory has been most completely developed by Stanley Mandelstam of the University of California at Berkeley and by Alexander M. Polyakov of the Landau Institute for Theoretical Physics near Moscow.

In superstring theory summing over histories requires summing over all the possible surfaces that join the initial and final states of a string or a set of interacting strings. The different paths can be thought of as fluctuations of the world sheet, which are just like the random, shimmering motions of a soap film at any temperature above absolute zero. As shown in Figure 11.5, a point particle is pictured as moving simultaneously along all possible world lines in space-time; a weight, or statistical probabil-

ity, is assigned to each path in such a way that the shortest paths in space-time are by far the most probable. The result is a collection of world lines that are densest along the classical trajectory (b). The cross section of the world lines through time, indicated by the broken vertical line, is shown in a. The density of the world lines through an arbitrary region of the cross section represents the probability of finding the string in that region at the corresponding time. Similarly, the quantum mechanics of a string is obtained by considering all possible world sheets with the same initial state and final state, weighted in such a way that the ones of smallest area are by far the most probable (d). The density of the closed loops in a given region of the temporal cross section of the world sheet (c) represents the probability of finding the complete string in that region at the corresponding time.

An important quantum-mechanical constraint on string theories was formulated in 1972 by Richard

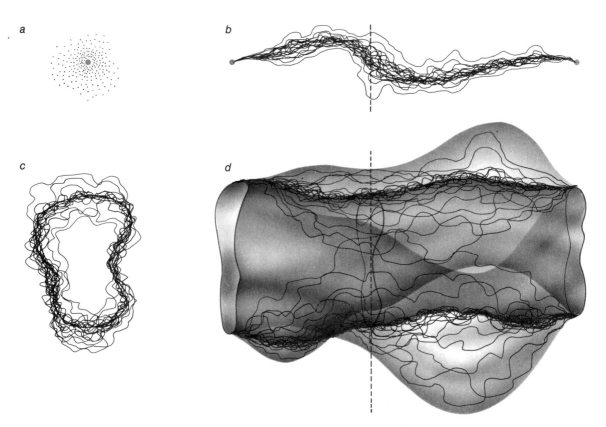

Figure 11.5 "SUMMING OVER HISTORIES" is one way to take account of the probabilistic interpretation of the path of a particle in quantum mechanics.

Brower of Boston University, Peter Goddard of the University of Cambridge and Charles B. Thorn of the University of Florida at Gainesville: Requiring symmetry under relabelings of the world-sheet coordinates of a free string undergoing quantum fluctuations is sufficient to determine the number of dimensions of the background space-time in which the string is moving. Remember that in the original dual-resonance theory the number of space-time dimensions is 26; in superstring theory 10 dimensions are required. This result had been intimated a year earlier in work by Claud W. V. Lovelace of Rutgers University. The constraint is a striking instance of how the formulation of the theory in terms of the two-dimensional world sheet leads to a rich and highly determined structure for the space in which the string is moving.

The vibrational frequencies of a superstring are determined by its tension, which is measured in units of energy per unit length, or mass squared in fundamental units. Since the theory is to describe gravity, the string tension must be closely related to the Planck energy, which is the only dimensional parameter in a gravitational theory. The tension of the string must therefore be on the order of 10^{19} GeV squared; in more familiar units that is equal to a force of 10^{39} tons. The frequencies of the normal vibrational modes of the string are therefore separated by huge gaps: particles corresponding to the lowest vibrational state are massless, but particles corresponding to the next vibrational state have a mass roughly equal to the mass of a speck of dust, which is enormous for an elementary particle. For higher-frequency vibrations the corresponding masses increase without limit.

The significance of the massless states of superstring theory is that they include not only the graviton, the spin-1 gauge particles of other forces and the spin-0 and spin-1/2 particles, but also the gravitino, the spin-3/2 gauge particle associated with supergravity. Thus for energies below the Planck energy the massless particles of superstring theories are the same ones found in supergravity theories.

A quantum string theory is different from the quantum theory of a point particle in another important respect. Again consider the sum over histories of a single closed string: the sum includes all possible connected surfaces that can be stretched, twisted or otherwise smoothly deformed into a cylinder without tearing. All such surfaces are said to

be topologically equivalent to the cylinder, and they include surfaces with long tentacles (see Figure 11.6). Under certain assignments of the time coordinate on the world sheet, the tentacles can be interpreted as motions in which two closed strings join to form a new one or in which a new closed string breaks away from the original one and disappears into the vacuum.

Thus the string automatically interacts with the background space in which it moves, even though interactions are not explicitly included in the sum over histories. In contrast, the sum over histories of a single point particle does not include any information about interactions with the background space. Because of the nature of the interaction of a string with its background, even the motion of a single string can be consistently described only if the curvature of the background space is severely constrained. This result was originally implied by the work of Daniel Friedan of the University of Chicago, done in 1979.

To appreciate the significance of the finding one must understand that the 10 dimensions required in a consistent superstring theory were not initially assumed to be curved at all. The theory was first formulated under the simplifying assumption that all 10 dimensions are equivalent, that is, all of them are "flat." If any superstring theory is to make sense of physical observations, however, six spatial dimensions must be highly curved. If the sum over the histories of the world sheet is to be consistent, the six dimensions must be curled up in one of a few special ways. They are said to form a kind of space called a Calabi-Yau space (after Eugenio Calabi of the University of Pennsylvania and Shing-Tung Yau of the University of California at San Diego); they may also form a generalization of such a space called an orbifold. These spaces lead to a promising scheme for explaining the physics of the four observable dimensions.

Much of the interest in superstring theories follows from the rich structure that results by requiring the theory to be consistent. If the theory is to give a realistic quantum account of the Yang-Mills forces, there is an empirical constraint in addition to the requirement that there be only four observable dimensions: the theory must lead to the observed chirality, or handedness, of the weak force.

The weak force is responsible for radioactive decay such as beta decay, which is an important

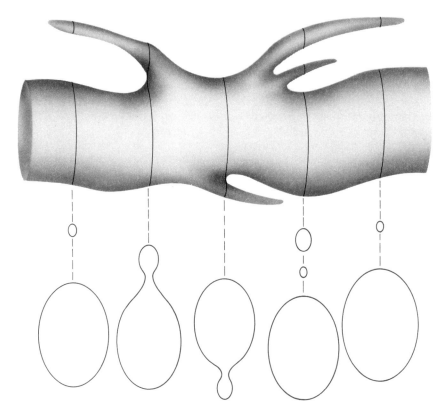

Figure 11.6 QUANTUM-MECHANICAL ANALYSIS of a single world sheet leads to strong constraints on the kind of curled-up background space in which the string is moving. The quantum-mechanical sum over histories includes configurations of the world sheet that describe strings materializing from the vacuum (*left*) or breaking away and disappearing into the vacuum (*right*). Thus string interactions with the background are automatically included in the quantum treatment of one string. The only background spaces known to be consistent with such effects are the so-called Calabi-Yau spaces or related spaces called orbifolds.

reaction in the sun. The force is chiral (from the Greek word *cheir*, meaning hand) in the sense that it gives rise to effects whose mirror-reflected counterparts do not exist in nature. Unfortunately the quantum-mechanical version of any chiral gauge theory is likely to violate conservation laws such as the conservation of electric charge. Such a violation is called a chiral anomaly; it signals a breakdown of gauge symmetry, which renders the theory inconsistent. Devising a theory that is chiral and yet avoids chiral anomalies is a delicate matter in four dimensions, and until recently it was thought to be impossible in 10 dimensions.

It is only when space has an odd number of dimensions (or when space-time has an even number) that the concept of chirality can be defined at all (see Figures 11.7 and 11.8). In any number of dimensions chirality depends on the outcome of successive mirror reflections along all the spatial axes. When space has an odd number of dimensions, mirror reflection along each spatial axis gives an odd number of reflections, and so a left-handed shape is transformed into a right-handed one. When space has an even number of dimensions, a reflection along each spatial axis leaves any shape unaltered: the reflected image can be rotated into the original shape. For example, the popular 11-dimensional theory of supergravity cannot lead to a chiral theory because it is formulated in 10 spatial dimensions (an even number). With nine spatial dimensions superstring theory can be chiral.

Even when the higher-dimensional theory is

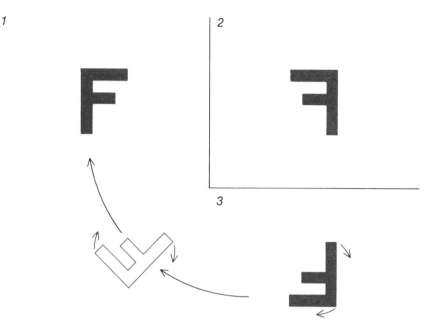

Figure 11.7 CHIRALITY, or handedness, cannot be defined in an even number of spatial dimensions (or in odd-dimensional space-time). For example, in two dimensions **successive reflections of the asymmetric letter F through both coordinate axes can be undone by a rotation.**

chiral, the process of curling up the extra dimensions generally washes out chirality. It is now thought that the observed chirality can be explained only if the chiral higher-dimensional theory is initially formulated to include the gauge symmetry group of a Yang-Mills force, in addition to the gravitational force. The symmetry group must be present in the flat, 10-dimensional theory before the possibility that the extra dimensions curl up can be considered.

In August, 1984, to the surprise of many physicists, Schwarz and I showed that a chiral theory free of anomalies can be formulated in 10 dimensions, provided the symmetry group of the Yang-Mills force is one of two special groups. The groups must be either $SO(32)$, a generalization to 32 internal dimensions of the symmetry of space-time, or the group $E_8 \times E_8$ (see Figure 11.9), the product of two exceptional, continuous groups that were first discovered by the French mathematician Elie Cartan. A third group known as $O(16) \times O(16)$, which also leads to freedom from anomalies, has recently been noted; its symmetry is a subsymmetry of the other two possible groups. The fact that the quantum consistency of a theory including gravity leads to an almost unique prediction of the unifying symmetry group was an exciting development. It has led to the current wave of enthusiasm for superstring theory.

The quantum mechanics of a single string is only an approximation to a full theory, not yet developed, in which the interactions of arbitrary numbers of strings would be described. Interactions arise when strings join or split. The probability that a given set of incoming strings leads to a given set of outgoing strings is determined by an infinite sequence of so-called Feynman diagrams. The diagrams describe all possible joinings and splittings of the world sheets, summed over histories for each case.

The simplicity of the sequence of possible Feynman diagrams for strings contrasts sharply with the complexity of the possible Feynman diagrams for point particles. Consider the possible diagrams for two interacting closed strings. First the two strings can join and then split apart. The Feynman diagram for the process is topologically equivalent to the surface of a sphere (see Figure 11.10). In the next term in the sequence the intermediate string splits into two strings, which then rejoin. The diagram is

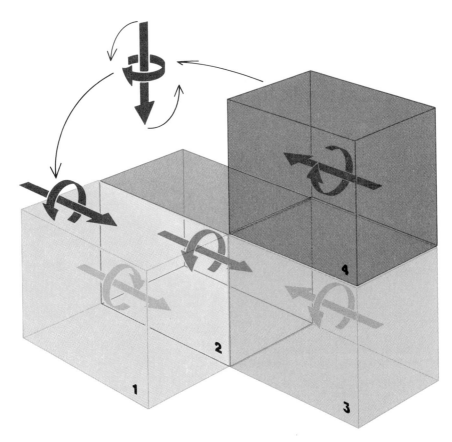

Figure 11.8 MIRROR REFLECTION through all the coordinate axes in a space with an odd number of dimensions (or an even-dimensional spacetime) gives rise to the possibility of finding chiral particles in the space. For example, a neutrino (1) travels at the speed of light and spins in a left-handed sense about the direction of its motion. Its mirror image after reflection through all three axes (4) spins in a right-handed sense; in other words, it is not equivalent to any rotated version of the original neutrino. Only the left-handed neutrino occurs in nature, which shows the laws of physics are asymmetric with respect to handedness.

topologically equivalent to a torus, or the surface of a doughnut. The sequence of diagrams is continued simply by adding holes to the doughnuts: two holes, three holes and so on.

Requiring symmetry under relabelings of coordinates on the torus introduces strong, new constraints to superstring theory. If the torus is cut in either one of two ways, twisted and glued back together, its topology is unchanged; the twisted coordinate system, however, cannot be continuously returned to the original one without cutting the surface again (see Figure 11.11). The invariance of the torus under such changes in coordinates leads to an important constraint in the heterotic superstring theory. A heterotic string can move through flat,

10-dimensional space-time only if the symmetry of the nongravitational forces in the theory is described either by the group $E_8 \times E_8$, the group *spin* $(32)/Z_2$, a variant of $SO(32)$, or the group $O(16) \times O(16)$.

The restriction to these groups is striking. Remember that Schwarz and I had earlier singled out the same groups by requiring the absence of chiral anomalies in any consistent 10-dimensional chiral gauge theory. The constraint that leads to these groups in the heterotic theory is related to remarkable mathematical properties of certain 16-dimensional lattices of points. Such lattices are constructed from the 16 internal dimensions I mentioned above.

There is another and almost equally striking fea-

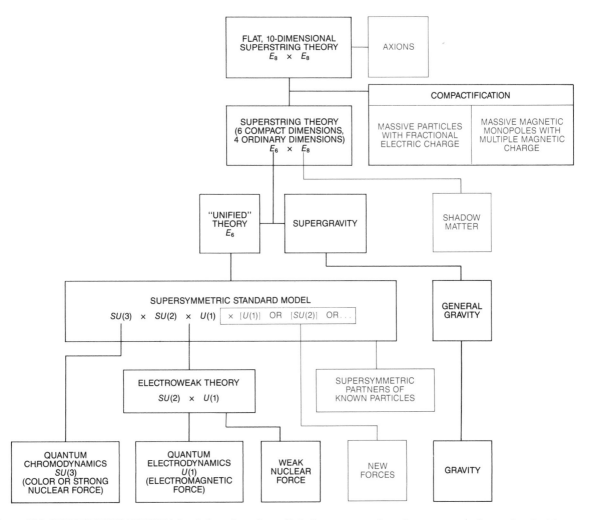

Figure 11.9 SUPERSTRING THEORY that is based on the huge symmetry group $E_8 \times E_8$ might make contact with observed physical phenomena by way of the connections shown in the logical diagram. When the effects of curvature in the six curled-up dimensions are considered, the theory resembles a supersymmetric grand-unified theory linked to supergravity, whose symmetry is associated with the group $E_6 \times E_8$. That symmetry can break down to give the standard model of elementary particles and forces, which accounts for the three nongravitational forces. New particles and forces that may be required by certain interpretations of the theory are shown in color.

ture of the torus diagram that indicates the profound difference between string theories and the theories of point particles. Among the many kinds of one-loop Feynman diagrams for point particles that are the analogues of the torus diagram, there are diagrams that give infinite answers when the sum over histories is calculated. Such infinities arise because the diagrams for point particles include the sum over histories when two interaction points are arbitrarily close in space-time. The contributions to the entire process of each one of an indefinitely large number of fluctuating paths must be taken into account, and that leads to infinite answers. In contrast, on a string Feynman diagram no one point can be identified as a splitting or joining point of two strings (see Figure 11.12). Hence, at least in the one-loop diagram of string interactions, the concept of an interaction point does not arise. The sum over

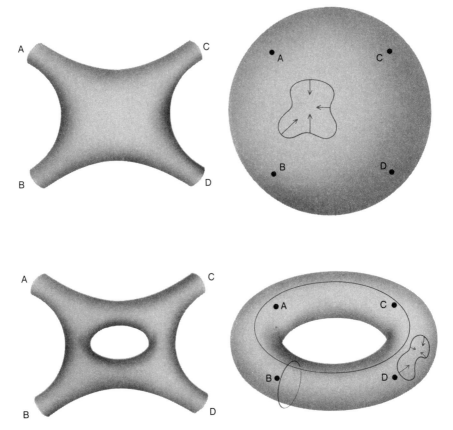

Figure 11.10 INTERACTIONS AMONG STRINGS are represented by world sheets topologically equivalent to a sphere or to a doughnut with an arbitrary number of holes. The surfaces are continuous except at the points that represent the incoming or outgoing strings (A, B, C, D). The sphere is distinct in that any closed curve on its surface can be continuously shrunk to a point. Two kinds of closed curves on the torus cannot be continuously shrunk to a point.

histories leads to finite answers in precisely those superstring theories whose Yang-Mills forces are associated with the symmetry groups $E_8 \times E_8$ or $SO(32)$.

An unfortunate feature of any quantum theory of gravity, whose natural energy scale is necessarily enormous, is the difficulty of extracting testable or observable predictions at more modest energies. For example, although superstring theory is initially formulated in 10 flat space-time dimensions with huge unifying symmetries, it is only at an inconceivably small distance scale or, equivalently, at extremely high energy or temperature that the curvature of the extra six dimensions might be negligible and the full symmetry of the theory would be realized. Thus any attempt to derive low-energy consequences of the theory must be considered speculative, and there are severe problems in obtaining concrete predictions. Nevertheless, it is now possible to give a plausible picture of how superstring theory might make contact with the phenomena observed in accelerators.

Many aspects of this picture are based on topological arguments, whose application to physics has been pioneered by Edward Witten of Princeton. Witten noted that the curling up of the extra dimensions and symmetry breaking go hand in hand: there must be a strong correlation between the curvature of the extra dimensions and the way in which the huge symmetry associated with the Yang-Mills forces breaks down to a smaller sym-

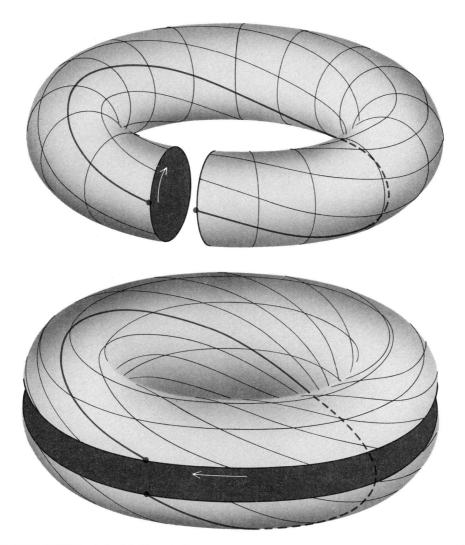

Figure 11.11 INVARIANCE of the labeling of the space and time coordinates on a world sheet with the topology of the torus is an even stronger requirement than it is on the sphere. A torus can be cut like a sausage (*upper diagram*) or like a yo-yo (*lower diagram*). One side of the cut can be given a full twist with respect to the other side, and the two boundaries can then be "glued" back together. The points on the surface are thereby labeled according to a set of coordinates (*black and colored curves*) that cannot be continuously deformed to the original set of coordinates. The requirement that superstring theory not be sensitive to such relabeling leads to strong conditions on the symmetry groups associated with the nongravitational forces of the theory.

metry. In flat 10-dimensional space-time the average values of the fields associated with the $E_8 \times E_8$ or $SO(32)$ Yang-Mills forces are zero. Witten showed that if the curvature of the space is nonzero in certain dimensions, the average values of these fields in the curved space must be nonzero in the same dimensions.

A nonzero average value for a field signals a transition to a phase with less symmetry, just as the nonzero magnetization of a ferromagnet signals the transition to the less symmetric magnetized phase. Thus if six dimensions of the original flat space-time become highly curved, the Yang-Mills forces in the theory of the resulting curved space are unified by a subsymmetry of $E_8 \times E_8$ or $SO(32)$. This result is just what one would want.

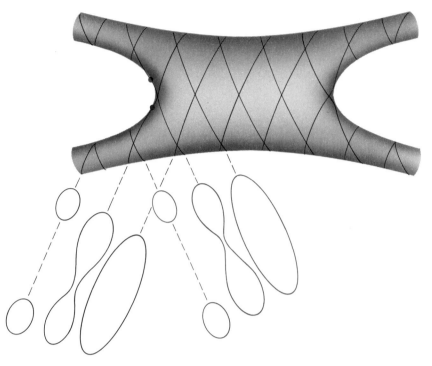

Figure 11.12 PRECISE INSTANT at which two strings join cannot be objectively specified because the definition of the time coordinate of a point on the world sheet depends on the observer. The colored coordinate lines indicate contours of equal time as defined by one observer. To this observer the two incoming strings appear to join at the colored point. To another observer the black lines are contours of equal time; the strings appear to join at the black point.

So far it has not been proved that the six spatial dimensions must be curled up in superstring theory. As I stated above, however, the requirement that the theory retain its large symmetry under relabelings of the coordinates of the world sheet forces the curled-up dimensions to form a Calabi-Yau space or perhaps an orbifold. Assuming six dimensions are curled up in this way, Philip Candelas of the University of Texas at Austin, Gary T. Horowitz of the University of California at Santa Barbara, Andrew E. Strominger of the Institute for Advanced Study in Princeton and Witten have devised an imaginative scheme for understanding how superstring theory might relate to phenomena measured at the energies of present experiments.

In the scheme of Candelas and his coworkers the average value of the electric field associated with the group $E_8 \times E_8$ is set equal to the nonzero value of the curvature of the six-dimensional space. Be-cause of special features of Calabi-Yau spaces (or the associated orbifolds), the symmetry breaks down in a special way to the group $E_6 \times E_8$, while still preserving its supersymmetry. It is reassuring to derive the group E_6 in this way because it is one of the groups studied in efforts to unify the Yang-Mills forces (see Figure 11.9).

The extra group factor E_8 enlarges the symmetry enormously, but it does not directly affect the observed particles. Particles carrying the forces associated with the extra E_8 symmetry are neutral with respect to the observed Yang-Mills forces, but they should exert gravitational attraction. It has been speculated that matter made up of such particles, which has been called shadow matter, might account for part of the unseen mass known to be present in the universe.

One immediate prediction of this interpretation of superstring theory is the existence of a particle called the axion. Such a particle has been postulated in order to avoid a violation of an important symmetry in QCD, the theory of the strong interaction. Although the axion has not yet been found, several groups of investigators are planning experiments that could detect it.

Many other physical consequences follow from purely topological properties of the six curled-up dimensions. For example, one of the most basic topological properties of a space is its Euler number. If the curled-up space were two-dimensional, the Euler number would be 2 minus twice the genus of the surface, where the genus is the number of holes. In six dimensions the Euler number is less easy to describe.

The observed fermions are naturally grouped into "generations" of quarks and leptons (see Figure 11.1). (The leptons are fermions that do not strongly interact.) Three generations have been discovered so far, and there is probably at most one more. The predicted number of generations is half the Euler number of the six-dimensional compact space. Because only a few of the known Calabi-Yau spaces and orbifold spaces have a small Euler number, only a few such spaces are candidates for describing observed physical phenomena.

The six-dimensional spaces known to have a small Euler number are spaces that also have "holes" in them (see Figure 11.13). The presence of the holes allows the fields associated with the symmetry group E_6 to be trapped. The trapping of the fields causes the symmetry to break down further, without the need for the massive so-called Higgs particles usually associated with this symmetry breaking. Quantum chromodynamics and the electroweak theory of standard elementary-particle physics, whose symmetries are described by the group product $SU(3) \times SU(2) \times U(1)$, can readily be obtained from this latest symmetry breaking. It is likely that the same symmetry breaking will also give rise to additional forces associated with remnants of E_6.

There are many ways a closed string can get trapped in the curled-up space. For example, the string can wind through a hole; its vibrations might then be manifest as massive particles with fractional electric charge or massive magnetic monopoles with multiple magnetic charges.

Although superstring theory has already opened many lines of investigation in both physics and mathematics, there are fundamental questions that cannot be answered until more is

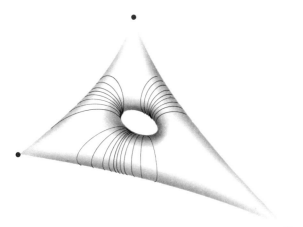
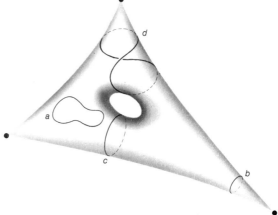

Figure 11.13 ORBIFOLD SPACE. A closed, two-dimensional surface has been stretched to form three sharp points called conical singularities. The surface is flat everywhere except at the singularities in the sense that it can be cut and laid out on a flat plane. The hole in the center of the surface indicates that the orbifold is not "simply connected": it can be cut through without falling into separate pieces. A closed path looping through a hole can be unwound after a fixed number of loops. Such holes cause lines of electromagnetic potential associated with the symmetry group E_6 to become trapped (*colored curves at left*). A closed string (*right*) can move freely in an orbifold (*a*), but it can also become trapped if it loops around a singularity where the curvature prevents it from unlooping (*b*). It loops through the hole (*c*) or wind first through the hole and then around a conical singularity (*d*).

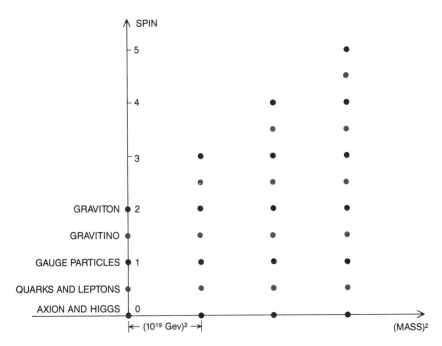

Figure 11.14 SPECTRUM OF STRING STATES is plotted for the heterotic string theory in which the extra six dimensions of space-time have been curled up. Each black dot represents a set of bosons and each colored dot a set of fermions. All string states that correspond to known particles are massless states; the states with nonzero mass form an infinite series whose masses are a whole number times the square of the Planck mass, which is 10^{19} GeV. For each mass the number of fermion states is equal to the number of boson states. If each possible spin direction is counted as a different state, there are 8,064 massless states, and 18,883,584 states at the first mass level; the number increases exponentially thereafter.

understood about the structure of the theory. For example, why is observed space-time approximately flat and four-dimensional? Can one prove that six dimensions are curled up? More to the point, can one explain why the four familiar dimensions are so large? Just after the big bang, when the size of the observed universe was on the order of the Planck distance, 10^{-35} meter, all 10 dimensions must have been curled up. During the subsequent expansion of the universe all but six of the dimensions must have begun to unfold and expand. A consistent account of the process might lead to the prediction of an observable cosmic remnant.

These issues are closely related to another fundamental question: Why is the cosmological constant so close to zero? This constant describes the part of the curvature of the universe that is not caused by matter; its value has been determined to be zero to within one part in 10^{120}, which makes it the most accurate measurement in all science. If superstring theory can account for the value, the explanation would be a convincing test of the theory.

As I mentioned above, the natural mass scale of superstring theory is so large that the masses of the heaviest particles observable in current accelerators are infinitesimal by comparison. As long as the supersymmetry included in superstring theories remains unbroken, the masses of the particles we observe are zero. The small, nonzero values of particles masses are therefore correlated with the breaking of supersymmetry at the relatively low temperatures of the world we observe. Nevertheless, there are strong arguments that the breaking of supersymmetry, and hence the precise values for the masses of particles observed in high-energy accelerators, cannot be understood in superstring theory as it is currently formulated.

In order to resolve such issues it is presumably necessary to understand the deep principles on which the theory is based. In a sense the development of superstring theories is in sharp contrast to the development of general relativity. In general relativity the detailed structure of the theory follows from Einstein's penetrating insights into the logic of the laws of physics. In superstring theories certain details have come first; we are still groping for a unifying insight into the logic of the theory. For example, the occurrence of the massless graviton and gauge particles that emerge from superstring theories appears accidental and somewhat mysterious; one would like them to emerge naturally in a theory after the unifying principles are well established.

How can the logic of superstring theory be discovered? The principles of general relativity must be a special case of the more general principles of superstring theory, and so in a sense general relativity can serve as a guide. For example, by developing an analogy with general relativity I explained above how superstring theories are expected to extend the idea of ordinary space-time to the space of all possible configurations of a string. This idea is under intensive investigation. Even more radical is the suggestion that the theory should be studied solely in its two-dimensional formulation; no reference at all would then be made to the coordinates of the space and time in which we live.

Whatever the resolution of these possibilities may be, developing a deep understanding of the logical status of the theory will undoubtedly lead to profound mathematical and physical problems. It should also lead to a better understanding of the predictions of superstring theory. The prospect is for a period of intellectual ferment and rapid advance.

The Structure of Quarks and Leptons

They have been considered the elementary particles of matter, but instead they may consist of still smaller entities confined within a volume less than a thousandth the size of a proton.

• • •

Haim Harari
April, 1983

I n the past 100 years the search for the ultimate constituents of matter has penetrated four layers of structure. All matter has been shown to consist of atoms. The atom itself has been found to have a dense nucleus surrounded by a cloud of electrons. The nucleus in turn has been broken down into its component protons and neutrons. More recently it has become apparent that the proton and the neutron are also composite particles; they are made up of the smaller entities called quarks. What comes next? It is entirely possible that the progression of orbs within orbs has at last reached an end and that quarks cannot be more finely divided. The leptons, the class of particles that includes the electron, could also be elementary and indivisible. Some physicists, however, are not at all sure the innermost kernel of matter has been exposed. They have begun to wonder whether the quarks and leptons too might not have some internal composition.

The main impetus for considering still another layer of structure is the conviction (or perhaps prejudice) that there should be only a few fundamental building blocks of matter. Economy of means has long been a guiding principle of physics, and it has served well up to now. The list of the basic constitu-

ents of matter first grew implausibly long toward the end of the 19th century, when the number of chemical elements, and hence the number of species of atoms, was approaching 100. The resolution of atomic structure solved the problem, and in about

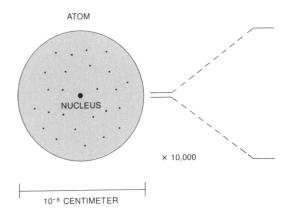

Figure 12.1 HIERARCHY OF PARTICLES in the structure of matter currently has four levels. All matter is made up of atoms; the atom consists of a nucleus surrounded by electrons; the nucleus is composed of protons and neu-

1935 the number of elementary particles stood at four: the proton, the neutron, the electron and the neutrino. This parsimonius view of the world was spoiled in the 1950's and 1960's; it turned out that the proton and the neutron are representatives of a very large family of particles, the family now called hadrons. By the mid-1960's the number of fundamental forms of matter was again roughly 100. This time it was the quark model that brought relief. In the initial formulation of the model all hadrons could be explained as combinations of just three kinds of quarks.

Now it is the quarks and leptons themselves whose proliferation is beginning to stir interest in the possibility of a simpler scheme. Whereas the original model had three quarks, there are now thought to be at least 18, as well as six leptons and a dozen other particles that act as carriers of forces. Three dozen basic units of matter are too many for the taste of some physicists, and there is no assurance that more quarks and leptons will not be discovered. Postulating a still deeper level of organization is perhaps the most straightforward way to reduce the roster. All the quarks and leptons would then be composite objects, just as atoms and hadrons are, and would owe their variety to the number of ways a few smaller constituents can be brought together. The currently observed diversity of nature would be not intrinsic but combinatorial.

It should be emphasized that as yet there is no evidence quarks and leptons have an internal structure of any kind. In the case of the leptons, experiments have probed to within 10^{-16} centimeter and found nothing to contradict the assumption that leptons are pointlike and structureless. As for the quarks, it has not been possible to examine a quark in isolation, much less to discern any possible internal features. Even as a strictly theoretical conception, the subparticle idea has run into difficulty: no one has been able to devise a consistent description of how the subparticles might move inside a quark or a lepton and how they might interact with one another. They would have to be almost unimaginably small: if an atom were magnified to the size of the earth, its innermost constituents could be no larger than a grapefruit. Nevertheless, models of quark and lepton substructure make a powerful appeal to the aesthetic sense and to the imagination: they suggest a way of building a complex world out of a few simple parts.

A ny theory of the elementary particles of matter must also take into account the forces that act between them and the laws of nature that govern the forces (see Figure 12.2). Little would be gained in simplifying the spectrum of particles if the number of forces and laws were thereby increased. As it happens, there has been a subtle interplay between the list of particles and the list of forces throughout the history of physics.

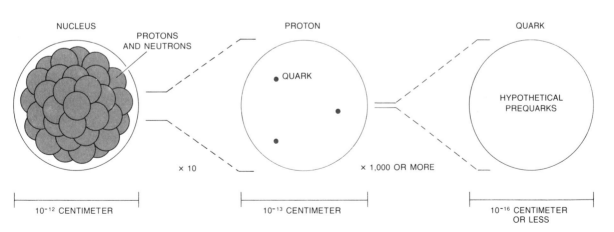

trons; each proton and neutron is thought to be composed of three quarks. Recent speculations might add a fifth level: the quark might be a composite of hypothetical finer constituents, which can be generically called prequarks. The leptons, the class of particles that includes the electron, could also consist of prequarks.

In about 1800 four forces were thought to be fundamental: gravitation, electricity, magnetism and the short-range force between molecules that is responsible for the cohesion of matter. A series of remarkable experimental and theoretical discoveries then led to the recognition that electricity and magnetism are actually two manifestations of the same basic force, which was soon given the name electromagnetism. The discovery of atomic structure brought a further revision. Although an atom is electrically neutral overall, its constituents are charged, and the short-range molecular force came to be understood as a complicated residual effect of electromagnetic interactions of positive nuclei and negative electrons. When two neutral atoms are far apart, there are practically no electromagnetic forces between them. When they are near each other, however, the charged constituents of one atom are able to "see" and influence the inner charges of the other, leading to various short-range attractions and repulsions.

As a result of these developments physics was left with only two basic forces. The unification of electricity and magnetism had reduced the number by one, and the molecular interaction had been demoted from the rank of a fundamental force to that of a derivative one. The two remaining fundamental forces, gravitation and electromagnetism, were both long-range. The exploration of nuclear structure, however, soon introduced two new short-range forces. The strong force binds protons and neutrons together in the nucleus, and the weak force mediates certain transformations of one particle into another, as in the beta decay of a radioactive nucleus. Thus there were again four forces.

The development of the quark model and the accompanying theory of quark interactions was the next occasion for revising the list of forces. The quarks in a proton or a neutron are thought to be held together by a new long-range fundamental force called the color force, which acts on the quarks because they bear a new kind of charge called color. (Neither the force nor the charge has any relation to ordinary colors.) Just as an atom is made up of electrically charged constituents but is itself neutral, so a proton or a neutron is made up of colored quarks but is itself colorless. When two colorless protons are far apart, there are essentially no color forces between them, but when they are near, the colored quarks in one proton "see" the color charges in the other proton. The short-range attractions and repulsions that result have been identified

with the effects of the strong force. In other words, just as the short-range molecular force became a residue of the long-range electromagnetic force, so the short-range strong force has become a residue of the long-range color force.

One more chapter can be added to this abbreviated history of the forces of nature. A deep and beautiful connection has been found between electromagnetism and the weak force, bringing them almost to the point of full unification. They are clearly related, but the connection is not quite as close as it is in the case of electricity and magnetism, and so they must still be counted as separate forces. Therefore the current list of fundamental forces still has four entries: the long-range gravitational, electromagnetic and color forces and the short-range weak force. Within the limits of present knowledge all natural phenomena can be understood through these forces and their residual effects.

The evolution of ideas about particles and that of ideas about forces are clearly interdependent. As new basic particles are found, old ones turn out to be composite objects. As new forces are discovered, old ones are unified or reduced to residual status. The lists of particles and forces are revised from time to time as matter is explored at smaller scale and as theoretical understanding progresses. Any change in one list inevitably leads to a modification of the other. The recent speculations about quark and lepton structure are no exception; they too call for changes in the complement of forces. Whether the changes represent a simplification remains to be seen.

Of the four established fundamental forces, gravitation must be put in a category apart. It is too feeble even to be detected in the interactions of individual particles, and it is not understood in terms of microscopic events. For the other three forces successful theories have been developed and are now widely accepted. The three theories are distinct, but they are consistent with one another; taken together they constitute a comprehensive model of elementary particles and their interactions, which I shall refer to as the standard model.

In the standard model the indivisible constituents of matter are the quarks and the leptons (see Figure 12.3). It is convenient to discuss the leptons first. There are six of them: the electron and its companion the electron-type neutrino, the muon and the muon-type neutrino and the tau and the tau-type neutrino. The electron, the muon and the tau have

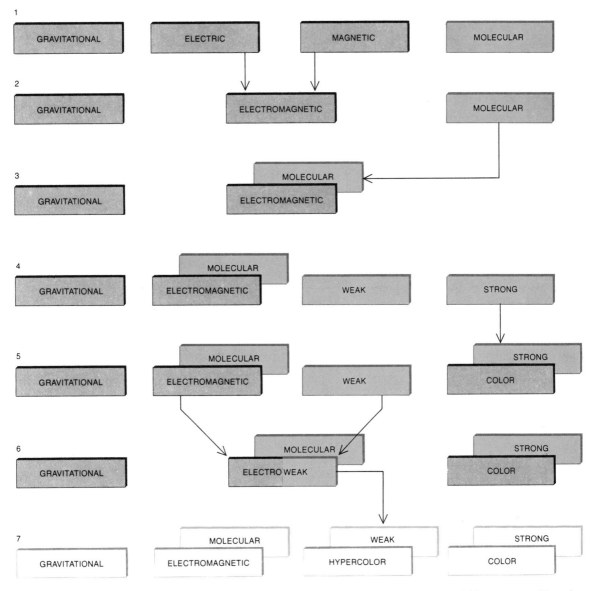

Figure 12.2 THE SCHEME OF CLASSIFICATION for fundamental forces of nature evolved together with the list of elementary particles. Long-range forces are shown in gray and short-range ones in color. Maxwell unified electricity and magnetism, and with the discovery of atomic structure it became apparent that the molecular force is a residual effect of electromagnetism. The discovery of the atomic nucleus introduced two short-range forces, the weak and the strong, which in the quark model becomes a residue of a new long-range force called the color force. Furthermore, a deep relation has been found between the weak force and electromagnetism, so that they can be considered partially unified. The sixth row of the chart represents the forces of nature as they are now understood in the "standard model" of elementary-particle physics.

an electric charge of −1; the three neutrinos are electrically neutral.

There are also six basic kinds of quark, which have been given the names up, down, charmed, strange, top and bottom, or u, d, c, s, t and b. (The top quark has not yet been detected experimentally, and neither has the tau-type neutrino, but few theorists doubt their existence.) The u, c and t quarks have an electric charge of +2/3, the d, s and b quarks a charge of −1/3. In addition each quark type has three possible colors, which I shall designate red, yellow and blue. Thus if each colored quark is counted as a separate particle, there are 18 quark varieties altogether. Note that each quark carries both color and electric charge, but none of the leptons are colored.

For each particle in this scheme there is an antiparticle with the same mass but with opposite values of electric charge and color. The antiparticle of the electron is the positron, which has a charge of +1. The antiparticle of a red u quark, with a charge of +2/3, is an antired $\bar{u}$ antiquark, with a charge of −2/3.

The color property of the quarks is analogous in many ways to electric charge, but because there are three possible colors it is appreciably more complicated. Electrically charged particles can be brought together to form an electrically neutral system in only one way: by combining equal quantities of positive and negative charge. A colorless composite particle can be formed out of colored quarks in much the same way, namely by combining a colored quark and an anticolored antiquark. In the case of color, however, there is a second way to form a neutral state: any composite system with equal quantities of all three colors or of all three anticolors is also colorless. For this reason a proton consisting of one red quark, one yellow quark and one blue quark has no net color.

One further property of the quarks and leptons should be mentioned: each particle has a spin, or intrinsic angular momentum, equal to one-half the basic quantum-mechanical unit of angular momentum. When a particle with a spin of 1/2 moves along a straight line, its intrinsic rotation can be either clockwise or counterclockwise when the particle is viewed along the direction of motion. If the spin is clockwise, the particle is said to be right-handed, because when the fingers of the right hand curl in the same direction as the spin, the thumb indicates the direction of motion. For a particle with the opposite sense of spin a left-hand rule describes the motion, and so the particle is said to be left-handed.

In the standard model the three forces that act on the quarks and leptons are described by essentially the same mathematical structure. It is known as a gauge-invariant field theory or simply a gauge theory. Each force is transmitted from one particle to another by carrier fields, which in turn are embodied in carrier particles, or gauge bosons.

The gauge theory of the electromagnetic force, called quantum electrodynamics or QED, is the earliest and simplest of the three theories. It was devised in the 1940's by Richard P. Feynman, Julian S. Schwinger and Sin-Itiro Tomonaga. QED describes the interactions of electrically charged particles, most notably the electron and the positron. There is one kind of gauge boson to mediate the interactions; it is the photon, the familiar quantum of electromagnetic radiation, and it is massless and has no electric charge of its own. QED is probably the most accurately tested theory in physics. For example, it correctly predicts the magnetic moment of the electron to at least 10 significant digits.

The theory of the color force was formulated by analogy to QED and is called quantum chromodynamics or QCD. It was developed over a period of almost two decades through the efforts of many theoretical physicists. In QCD particles interact by virtue of their color rather than their electric charge. The gauge bosons of QCD, which are responsible for binding quarks inside a hadron, are called gluons. Like the photon, the gluons are massless, but whereas there is just one kind of photon, there are eight species of gluons. A further difference between the photon and the gluons turns out to be even more important. Although the photon is the intermediary of the electromagnetic force, it has no electric charge and hence gives rise to no electromagnetic forces of its own (or at least none of significant magnitude). The gluons, in contrast, are not colorless. They transmit the color force between quarks but they also have color of their own and respond to the color force. This reflexiveness, whereby the carrier of the force acts on itself, makes a complete mathematical analysis of the color force exceedingly difficult.

One peculiarity that seems to be inherent in QCD is the phenomenon of color confinement. It is thought that the color force somehow traps colored objects (such as quarks and gluons) inside composite objects that are invariably colorless (such as protons and neutrons). The colored particles can never escape (although they can form new colorless combinations). It is because of color confinement, physicists suppose, that a quark or a gluon has never been seen in isolation. I must stress that although the idea of color confinement is now widely accepted, it has not been proved to follow from QCD. There may still be surprises in store.

The weak force is somewhat different from the other two, but it can nonetheless be described by a

		QUARKS		LEPTONS	
GENERATIONS	FIRST	u (UP)	d (DOWN)	e (ELECTRON)	ν_e (ELECTRON-TYPE NEUTRINO)
	SECOND	c (CHARMED)	s (STRANGE)	μ (MUON)	ν_μ (MUON-TYPE NEUTRINO)
	THIRD	t (TOP)	b (BOTTOM)	τ (TAU)	ν_τ (TAU-TYPE NEUTRINO)

CHARGES		QUARKS		LEPTONS	
	ELECTRIC	+2/3	−1/3	−1	0
	COLOR	RED / YELLOW / BLUE	RED / YELLOW / BLUE	COLORLESS	COLORLESS

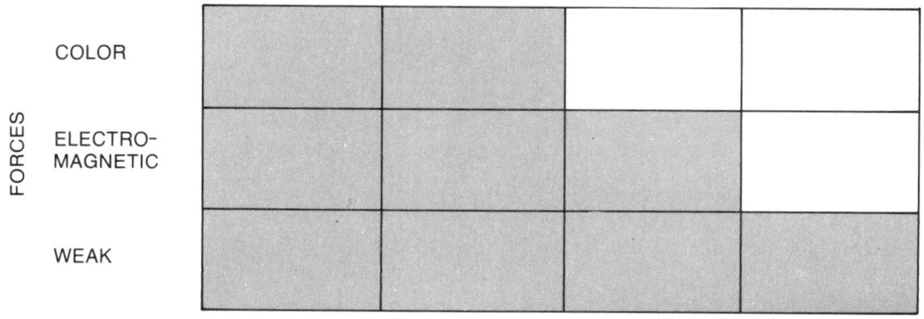

FORCES					
	COLOR				
	ELECTRO-MAGNETIC				
	WEAK				

Figure 12.3 STANDARD MODEL of elementary particles includes three "generations" of quarks and leptons, although all ordinary matter can be constructed out of the particles of the first generation alone. The quarks are distinguished by fractional values of electric charge and by a property that is fancifully called color: each quark type comes in red, yellow and blue versions. The leptons have integer units of electric charge and are colorless. The two classes of particles also differ in their response to the various forces. Only the quarks are subject to the color force, and as a result they may be permanently confined inside composite particles such as the proton.

gauge theory of the same general kind. The theory was worked out, and the important connection between the weak force and electromagnetism was established, in the 1960's and early 1970's by a large number of investigators. Notable contributions were made (in chronological order) by Sheldon Lee Glashow of Harvard University, Steven Weinberg of the University of Texas at Austin, Abdus Salam of the International Centre for Theoretical Physics in Trieste and Gerard 't Hooft of the University of Utrecht.

Curiously, the charges on which the weak force acts are associated with the handedness of a particle. Among both quarks and leptons left-handed particles and right-handed antiparticles have a weak charge, but right-handed particles and left-handed antiparticles are neutral with respect to the weak force. What is odder still, the weaker charge is not conserved in nature: a unit of charge can be created out of nothing or can disappear into the vacuum. In contrast, the net quantity of electric charge in an isolated system of particles can never be altered, and neither can the net color. The weak force is also distinguished by its exceedingly short range; its effects extend only to a distance of about 10^{-16} centimeter, or roughly a thousandth of the diameter of a proton.

In the gauge theory of the weak force both the failure of the weak charge to be conserved and the short range of the force are attributed to a mechanism called spontaneous symmetry breaking, which I shall discuss in greater detail below. For now it is sufficient to note that the symmetry-breaking mechanism implies that the weak charge, and the associated handedness of particles, should be conserved at extremely high energy, where a particle's mass is a negligible fraction of its kinetic energy.

Spontaneous symmetry breaking also requires that the gauge bosons of the weak force be massive particles; indeed, they have masses approximately 100 times the mass of the proton. In the standard model there are three such bosons: two of them, designated W^+ and W^-, carry electric charge as well as weak charge; the third, designated Z^0, is electrically neutral. The large mass of the weak bosons accounts for the short range of the force. According to the uncertainty principle of quantum mechanics, the range of a force is inversely proportional to the mass of the particle that transmits it. Thus electromagnetism and the color force, being carried by massless gauge bosons, are effectively infinite in range, whereas the weak force has an exceedingly small sphere of influence. Spontaneous symmetry breaking has still another consequence: it predicts the existence of at least one additional massive particle, separate from the weak bosons. It is called the Higgs particle after Peter Higgs of the University of Edinburgh, who made an important contribution to the theory of spontaneous symmetry breaking.

In the past 10 years the successes of the standard model have given physicists a good deal of self-confidence. All known forms of matter can be constructed out of the 18 colored quarks and the six leptons of the model. All observed interactions of matter can be explained as exchanges of the 12 gauge bosons included in the model: the photon, the eight gluons and the three weak bosons. The model seems to be internally consistent; no one part is in conflict with any other part, and all measurable quantities are predicted to have a plausible, finite value. Internal consistency is not a trivial achievement in a conceptual system of such wide scope. So far the model is also consistent with all experimental results, that is to say, no clear prediction of the

model has yet been contradicted by experiment. To be sure, there are some important predictions that have not yet been fully verified; most notably, the tau-type neutrino, the top quark, the weak bosons and the Higgs particle must be found. The first direct evidence of W bosons was recently reported by a group of experimenters at CERN, the European Laboratory for Particle Physics in Geneva. In the next several years new particle accelerators and more sensitive detecting apparatus will test the remaining predictions of the model. Most physicists are quite certain they will be confirmed.

If the standard model has proved so successful, why would anyone consider more elaborate theories? The primary motivation is not a suspicion that the standard model is wrong but rather a feeling that it is less than fully satisfying. Even if the model gives correct answers for all the questions it addresses, many questions are left unanswered and many regularities in nature remain coincidental or arbitrary. In short, the model itself stands in need of explanation.

The strongest hint of some organizing principle beyond the standard model is the proliferation of elementary particles. The known properties of matter are not so numerous or diverse that 24 particles are needed to represent them all. Indeed, there seems to be a great deal of repetition in the spectrum of quarks and leptons. There are three leptons with an electric charge of −1, three neutral leptons, three quarks with a charge of +2/3 and three quarks with a charge of −1/3. Everything is triplicated, and for no apparent reason. A world constructed by choosing one particle from each of the four groups would seem to have all the necessary variety.

As it turns out, all ordinary matter can indeed be formed from a subset that includes just the u quark, the d quark, the electron and the electron-type neutrino. These four particles and their antiparticles make up the "first generation" of quarks and leptons. The remaining quarks and leptons merely repeat the same pattern in two additional generations without seeming to add anything new. Corresponding particles in different generations are identical in all respects except one: they have different masses. The d, s and b quarks, for example, respond in precisely the same way to the electromagnetic, color and weak forces. For some unknown reason, however, the s quark is roughly 20 times as heavy as the d quark, and the b quark is approximately 600 times as heavy as the d. The mass ratios of the other quarks and of the charged leptons are likewise large and unexplained. (The masses of the neutrinos are too small to have been measured; it is not yet known whether the neutrinos are merely very light or are entirely massless.)

The presence of three generations of quarks and leptons begs for an explanation. Why does nature repeat itself? The pattern of particle masses is also mysterious. In the standard model the masses are determined by approximately 20 "free" parameters that can be assigned any values the theorist chooses; in practice the values are generally based on experimental findings. Is it possible the 20 parameters are all unrelated? Are they fundamental constants of nature with the same status as the velocity of light or the electric charge of the electron? Probably not.

A further tantalizing regularity can be perceived in the electric charges of the quarks and leptons: they are all related by simple ratios and are all integer multiples of one-third the electron charge. The standard model supplies no reason; in principle the charge ratios could have any values. It can be deduced from observation that the ratios of one-third and two-thirds that define the quark charges are not approximations. The proton consists of two u quarks and a d quark, with charges of $2/3 + 2/3 - 1/3$, or $+1$. If these values were not exact and the quarks instead had charges of, say, $+.617$ and $-.383$, the magnitude of the proton's charge would not be exactly equal to that of the electron's, and ordinary atoms would not be electrically neutral. Since atoms can be brought together in enor-

mous numbers, even a slight departure from neutrality could be readily detected.

If the particles and antiparticles that make up a single generation are arranged according to their charge (see Figure 12.4), it is found that every value from -1 to $+1$ in intervals of one-third is occupied by one particle (or, in the case of zero charge, by two particles, namely the neutrino and the antineutrino). The pattern formed raises still more questions. Why has nature favored these values of electric charge but no others, such as $+4/3$ or $-5/3$? It is apparent that all particles with integral charge are colorless and all those with fractional charge are colored. Is there some relation between the electric charge of a particle and its color or between the quarks and the leptons? The standard model implies no such relations, but they seem to exist.

Another motivation for looking beyond the standard model is the continuing desire to unify the fundamental forces, or at least to find some relation among them. The cause of parsimony would be served, for example, if two of the forces could be consolidated, as electricity and magnetism were, or if one force could be made a residue of another, as the strong force was made a residue of the color force. Ironically, it may turn out that a simplification of this kind can be attained only by introducing still more forces.

A theory that "goes beyond" the standard model need not contradict or invalidate it. The standard model may emerge as a very good approximation of the deeper theory. The standard model gives a remarkably successful description of all phenom-

	ELECTRIC CHARGE						
	+1	+2/3	+1/3	0	−1/3	−2/3	−1
ANTILEPTONS	$\bar{e}$			$\bar{\nu}_e$			
QUARKS		u			d		
ANTIQUARKS			$\bar{d}$			$\bar{u}$	
LEPTONS				ν_e			e

Figure 12.4 FIRST GENERATION of quarks and leptons forms an orderly pattern when the particles are arranged according to their electric charge. All values of charge from +1 to −1 in intervals of 1/3 are represented. All colored particles have fractional charge and all colorless ones have integral charge. The pattern is an arbitrary feature of the standard model, where charge and color are independent, but it might have some explanation if quarks and leptons are composite.

ena at distances no smaller than about 10^{-16} centimeter. A deeper theory should therefore focus on events at a still smaller scale. If there are new constituents to be discovered, they must exist within such minuscule regions of space. If there are new forces, their action must be effective only at a distance of less than 10^{-16} centimeter, either because the force is inherently short-range (following the example of the weak force) or because it is subject to some form of confinement (as the color force is).

The search for a theory beyond the standard model was launched almost 10 years ago, and by now several directions have been explored. One promising direction has led to the models known as grand unified theories, which incorporate the electromagnetic, color and weak forces into one fundamental force. The essential idea is to put all the quarks and leptons that make up one generation into a single family; new gauge bosons are then postulated to mediate interactions between the colored quarks and the colorless leptons. The theories account for the regularities noted in the distribution of electric charge and explain the exact commensurability of the quark and lepton charges. On the other hand, they do nothing to reduce the number of fundamental constants, they shed no light on the triplication of the generations and they create certain new theoretical difficulties of their own.

There have been several variations on the theme of grand unification. For example, the concept of horizontal symmetry tackles the triplication problem by establishing a symmetry relation among the generations. The mathematically beautiful idea called supersymmetry relates particles whose spin angular momentum is a half-integer (such as the quarks and leptons) to those with integer spin (such as the gauge bosons). The technicolor theory suggests that the Higgs particle of the standard model is a composite object made up of new fundamental entities; they would be bound together by a new force analogous to the color force and called technicolor. Each of these ideas answers some of the questions that remain open in the standard model. Each idea also fails to answer other questions, raises new difficulties and worsens existing ones, for example by further increasing the number of unrelated arbitrary constants.

In all the above schemes for grand unification it is explicitly assumed that the quarks, the leptons, the photon, the gluons and the weak bosons are the truly fundamental particles of the ultimate theory of nature. The alternative of suggesting that the quarks

and leptons are themselves composite is in one sense the most conservative and the least original hypothesis. It is a strategy that has worked before, repeatedly, in going from the atom to the nucleus to the proton to the quark. In another sense the idea of quark and lepton substructure is a most radical proposal. The electron has now been studied for almost a century, and its pointlike nature has been established very well indeed. In the case of the neutrino, which may turn out to be entirely massless, it is even more difficult to imagine an internal structure. The assertion that these particles and the others like them are composites will clearly have to overcome formidable obstacles if it is to have any future.

Offsetting the difficulties of the undertaking are its potential rewards. A fully successful composite model might resolve all the questions left unsettled in the standard model. Such a hypothetical theory would begin by introducing a new set of elementary particles, which I shall refer to generically as prequarks. Ideally there would not be too many of them. Each quark and lepton in the standard model would be accounted for as a combination of prequarks, just as each hadron can be explained as a combination of quarks. The mass of a quark or a lepton would no longer be an arbitrary constant of nature; instead it would be determined by the masses of the constituent prequarks and by the strength of the force that binds the prequarks together. The exact ratios that relate the charge of a quark to that of a lepton would be explained in a similar way: both kinds of composite particles would derive their charges from those of the same constituent prequarks. The entire pattern of quarks and leptons within a generation would presumably reflect some simple rules for combining the prequarks.

The existence of multiple generations might also be explained in a natural way. The quarks and leptons in the higher generations might have an internal constitution similar to that of the corresponding particles of the first generation; the differences could be in the energy and the state of motion of the constituents. Thus the s and b quarks would be excited states of the d quark, and the muon and the tau lepton would be excited states of the electron. Similar excited states are known in all other composite systems, including atoms, nuclei and hadrons. For example, at least a dozen hadrons have been identified in experiments as excited states of the proton; they and the proton itself are all thought to have essentially the same quark composition, namely uud.

This imaginary, ideal prequark theory accomplishes everything one might ask of it except for unifying the fundamental forces. Even there some progress is conceivable, since a new force would very likely be introduced to bind the prequarks together; the new force might lead to a new understanding of how the known forces are related. Imagining what a successful model might be like, however, is not at all the same thing as actually constructing a realistic and internally consistent one. So far no one has done it.

What has been lacking is a satisfactory theory of prequark dynamics, a theory that would describe how the prequarks move inside a quark or a lepton and that would enable one to calculate the mass and total energy of the system. As I shall set forth below, there are fundamental obstacles to the formulation of such a theory, although I would submit that they are not insurmountable. In the meantime, lacking any persuasive account of prequark motions, theorists have nonetheless been exploring the combinatorial possibilities of the prequark idea, that is, they have been examining the ways quarks and leptons might be built up as specific combinations of finer constituents.

In the past few years several dozen composite models have been proposed; they can be classified in perhaps four or five main groups. No single model solves all problems, answers all questions and is widely accepted. It would be unfair to describe only one scheme, but it is impractical to enumerate them all. I shall present a few of the central ideas.

The first explicit model of quark and lepton substructure was proposed in 1974 by Jogesh C. Pati of the University of Maryland at College Park and Salam, who have since returned to the topic several times in collaboration with John Strathdee of the International Centre for Theoretical Physics. It was they who introduced the term prequark, which I have adopted here as a generic name for hypothetical subconstituents of all kinds. The specific elementary particles of the model devised by Pati and Salam I shall call preons, which is another term of their invention.

The rationale for the preon model (see Figure 12.5) begins with the observation that every quark and lepton can be identified unambiguously by listing just three of its properties: electric charge, color and generation number. These properties, then, suggest a straightforward way of organizing a set of constituent particles. Three families of preons are

	PREON	ELECTRIC CHARGE	COLOR	GENERATION NUMBER
FLAVONS	f_1	+1/2	COLORLESS	0
FLAVONS	f_2	−1/2	COLORLESS	0
CHROMONS	c_R	+1/6	RED	0
CHROMONS	c_Y	+1/6	YELLOW	0
CHROMONS	c_B	+1/6	BLUE	0
CHROMONS	c_0	−1/2	COLORLESS	0
SOMONS	s_1	0	COLORLESS	1
SOMONS	s_2	0	COLORLESS	2
SOMONS	s_3	0	COLORLESS	3

Figure 12.5 PREON MODEL assigns three properties of quarks and leptons to three groups of hypothetical constituents called flavons, chromons and somons. A quark or a lepton is formed by choosing one preon from each group. The flavons have the primary responsibility for determining electric charge, the chromons determine color and the somons determine generation number. Ideally each kind of preon would carry just one property, but some adjustment is needed to differentiate the fractional electric charges of the quarks from the integral charges of the leptons. In the version of the model shown here the chromons carry electric charge as well as color.

needed. In one family the preons carry electric charge, in another they carry color and in the third they have some property that determines generation number. A given quark or lepton is assembled by selecting exactly one preon from each family.

The preons that determine generation number are called somons, from the Greek *soma*, meaning body, because they have a dominant influence on the mass of the composite system. Since there are three generations of quarks and leptons, there must be three somons. The color of the composite system is determined by preons called chromons; there are four of them, one with the color red, one yellow, one blue and one colorless. The remaining family of preons, which is assigned the role of defining electric charge, needs to have only two members in order for every quark and lepton to be uniquely identified. These last preons have been given the name flavons, after flavor, the whimsical term for whatever property it is that distinguishes the *u*

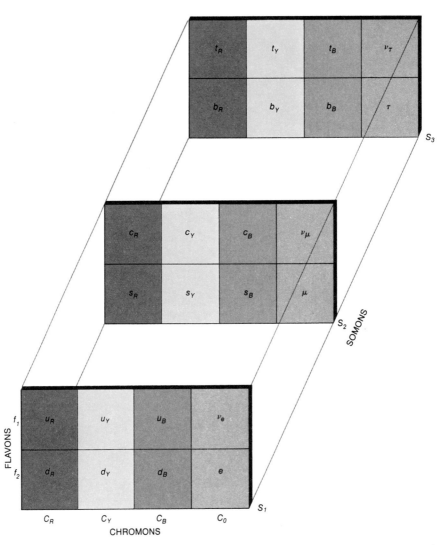

Figure 12.6 COMBINATIONS OF PREONS give rise to the 24 quarks and leptons of the three generations. For example, the red s quark is made up of somon S_2 (signify- ing that the composite is a second-generation particle) in combination with the red chromon and the negative flavon.

quark from the d quark, the c from the s, the neutrino from the electron and so on.

In the preon model the classification of a composite particle follows directly from its complement of preons. All leptons, for example, are distinguished by a colorless chromon, and all first-generation particles must obviously have a first-generation somon. In the allocation of electric charge, however, a complication arises. If there are only two flavons and if they are the sole carriers of electric charge, not all the charge values observed in nature can be reproduced. The u quark and the neutrino, for example, must have the same charge (because they include the same flavon), and so must the d quark and the electron. The problem can be solved in any of several ways. In one scheme electric charge is assigned to both the flavons and the chromons, and the total charge of a composite particle is equal to the sum of the two values. Models of this kind can be made to yield the correct charge states, but only by abandoning the principle of having each kind of preon carry just one property.

Another troublesome feature of the preon model is the requirement that composites be formed only by drawing one preon from each family. Why are there no particles made up of three chromons, say, or of two somons and a flavon? The exotic properties of such particles would make them quite conspicuous. It seems likely that if they existed, they would have been detected by now.

Many variations of the preon model have been proposed by other physicists, using the same basic idea but slightly different sets of preons. Notable among the variations are the models suggested by Hidezumi Terazawa, Yoichi Chikashige and Keiichi Akama of the University of Tokyo and by O. Wallace Greenberg and Joseph Sucher of the University of Maryland.

Perhaps the simplest model of quark and lepton structure is the rishon model, (see Figures 12.7 and 12.8) which I proposed in 1979. A similar idea was put forward at about the same time by Michael A. Shupe of the University of Illinois at Urbana-Champaign. The model has since been further developed and studied in great detail by Nathan Seiberg and me at the Weizmann Institute of Science in Rehovot. The model postulates just two species of fundamental building blocks, called rishons. (*Rishon* is the Hebrew adjective meaning first or primary.) One rishon has an electric charge of +1/3 and the other is electrically neutral. I designate them respectively T and V, for *Tohu Vavohu*, Hebrew for "formless and void," the description of the initial state of the universe given in the first chapter of Genesis.

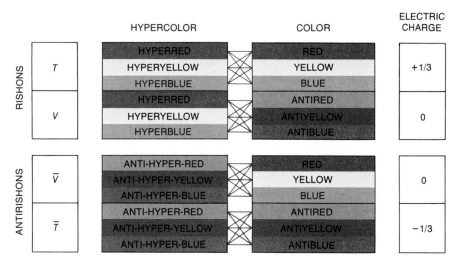

Figure 12.7 RISHON MODEL constructs all the quarks and leptons out of just two species of fundamental particles and their antiparticles. The rishons carry both hypercolor, a property associated with the force that binds them to one another, and ordinary color, which they convey to the composite systems they form. One rishon is electrically charged and the other is neutral.

The complementary antirishons have charges of $-1/3$ and zero and are designated $\overline{T}$ and $\overline{V}$.

The model has one simple rule for constructing a quark or a lepton: any three rishons can be assembled to form a composite system, or any three antirishons, but rishons and antirishons cannot be mixed in a single particle. The rule gives rise to 16 combinations, which reproduce exactly the properties of the 16 quarks, antiquarks, leptons and antileptons in the first generation. In other words, every quark and lepton in the first generation corresponds to some allowed combination of rishons or antirishons. (In this system of classification each color is counted separately.)

The pattern of quark and lepton charges is generated as follows. The TTT combination, with rishon charges of $1/3 + 1/3 + 1/3$, has a total charge of $+1$ and therefore corresponds to the positron; similarly, $\overline{T}\,\overline{T}\,\overline{T}$ has a total charge of -1 and is identified with the electron. The VVV and $\overline{V}\,\overline{V}\,\overline{V}$ combinations are both electrically neutral and represent the neutrino and the antineutrino respectively. The remaining allowed combinations yield fractionally charged quarks. TTV, with a charge of $+2/3$, is the u quark, and TVV, with a charge of $+1/3$, is the $\overline{d}$ antiquark. The analogous antirishon states $\overline{V}\,\overline{V}\,\overline{T}$ and $\overline{V}\,\overline{T}\,\overline{T}$ correspond to the d quark and the $\overline{u}$ antiquark.

The model also accounts successfully for the color of the composite systems. A T rishon can have any of the three colors red, yellow and blue, whereas a V rishon has an anticolor. Combinations such as $\overline{T}\,\overline{T}\,\overline{T}$ and VVV, which designate leptons, can be made colorless since they can include one rishon in each color or one in each anticolor. The other combinations, which yield quarks, must have a net color. For example, a TTV state might have the rishon colors red, blue and antiblue; the antiblue would cancel the blue, leaving the system with a net color of red. In this way the connection between color and electric charge, which was apparent but unexplained in the standard model, is readily understood. Because of the way electric charge and color are allotted to the rishons, all composite systems with fractional charge turn out to be colored, and all systems with an integer charge can be made colorless.

Other regularities of the standard model also lose their air of mystery when rishons are introduced. Consider the hydrogen atom, made up of a proton and an electron, or in terms of quarks and leptons two u quarks, a d quark and an electron. The total rishon content of the quarks is four T's, one $\overline{T}$, two

V's and two $\overline{V}$'s. The electric charge of the $\overline{T}$ cancels the charge of one T rishon, and the V's and $\overline{V}$'s also cancel (they have no charge in any case), leaving the proton with a net charge equal to that of a TTT system. The electron's rishon content is just the opposite: $\overline{T}\,\overline{T}\,\overline{T}$. Thus it is evident why the proton and the electron have charges of equal magnitude and why the hydrogen atom is neutral: the ultimate sources of the charge are pairs of matched particles and antiparticles.

The rishon model and many other models that explain the pattern of the first generation have difficulty accounting for the second and third generations. It would seem that such models lend themselves well to the scheme of forming each particle in the higher generations as an excited state of the corresponding particle in the first generation. The simplest idea would be to describe the muon, for example, as having the same prequark constituents as the electron, but in the muon the prequarks would have some higher-energy configuration. It is an elegant idea but, regrettably, it appears to be unworkable. The scheme implies differences in energy between the successive excited states that are much larger than the actual differences. The flaw is a fundamental one, and there seems to be no remedy.

Other possible mechanisms for creating multiple generations have been considered. Several physicists have suggested that the higher-generation relatives of a given state might be created by adding a Higgs particle, the "extra" particle associated with the weak bosons in the standard model. Because a Higgs particle has no electric charge or color or even spin angular momentum, adding one to a composite system would alter only the mass. Hence an electron might be converted into a muon by adding one Higgs particle or into a tau by adding two or more Higgs particles. Seiberg and I have proposed another possible mechanism: a higher-generation particle could be formed by the addition of pairs of prequarks and antiprequarks. All charges and other properties must cancel in such a pair, and so again only the mass would be affected.

These ideas are currently at the stage of unrestrained speculation. No one knows what distinguishes the three generations from one another, or why there are three or whether there may be more. No explanation can be given of the mass difference between the generations. In short, the triplication of the generations is still a major unsolved puzzle.

RISHON COMBINATION	PARTICLE	COLOR	ELECTRIC CHARGE
TTT	$\bar{e}$	COLORLESS	+1
TTV	u	RED / YELLOW / BLUE	+2/3
TVV	$\bar{d}$	ANTIRED / ANTIYELLOW / ANTIBLUE	+1/3
VVV	ν_e	COLORLESS	0

RISHON COMBINATION	PARTICLE	COLOR	ELECTRIC CHARGE
$\overline{VVV}$	$\bar{\nu}_e$	COLORLESS	0
$\overline{VVT}$	d	RED / YELLOW / BLUE	−1/3
$\overline{VTT}$	$\bar{u}$	ANTIRED / ANTIYELLOW / ANTIBLUE	−2/3
$\overline{TTT}$	e	COLORLESS	−1

Figure 12.8 COMBINATIONS OF RISHONS taken three at a time give a correct accounting of all the quarks and leptons (and antiquarks and antileptons) in any one generation. The pattern of electric charges noted in the standard model, and the apparent relation between fractional charge and color, emerge as natural consequences of the way the rishons combine. All the allowed combinations of three rishons or of three antirishons are neutral with respect to hypercolor.

A third kind of substructure model deserves mention. It tries to relate the possibility of quark and lepton structure to another fundamental problem: understanding the relativistic quantum theory of gravitation. Ideas of this kind have been explored by John Ellis, Mary K. Gaillard, Luciano Maiani and Bruno Zumino of CERN. One approach to their ideas is to consider the distances at which prequarks interact: the experimental limit is less than 10^{-16} centimeter, but the actual distance could be several orders of magnitude smaller still. At about 10^{-34} centimeter the gravitational force becomes strong enough to have a significant effect on individual particles. If the scale of the prequark interactions is this small, gravitation cannot be neglected. Ellis,

Gaillard, Maiani and Zumino have outlined an ambitious program that aims to unify all the forces, including gravitation, in a scheme that treats not only the quarks and leptons but also the gauge bosons as composite particles. Like other composite models, however, this one has serious flaws.

Any prequark model, regardless of its details, must supply some mechanism for binding the prequarks together. There must be a powerful attractive force between them. One strategy is to postulate a new fundamental force of nature analogous in its workings to be color force of the standard model. To emphasize the analogy the new force is called the hypercolor force and the carrier fields are called hypergluons. The prequarks are assumed to have hypercolor, but they combine to form hypercolorless composite systems, just as quarks have ordinary color but combine to form colorless protons and neutrons. The hypercolor force presumably also gives rise to the property of confinement, again in analogy to the color force. Hence all hypercolored prequarks would be trapped inside composite particles, which would explain why free prequarks are not seen in experiments. An idea of this kind was first proposed by 't Hooft, who studied some of its mathematical implications but also expressed doubt that nature actually follows such a path.

The typical radius of hypercolor confinement must be less than 10^{-16} centimeter. Only when matter is probed at distances smaller than this would it be possible to see the hypothetical prequarks and their hypercolors. At a range of 10^{-14} or 10^{-15} centimeter hypercolor effectively disappears; the only objects visible at this scale of resolution (the quarks and leptons) are neutral with respect to hypercolor. At a range of 10^{-13} centimeter ordinary color likewise fades away, and the world seems to be made up entirely of objects that lack both color and hypercolor: protons, neutrons, electrons and so on.

The notion of hypercolor is well suited to a variety of prequark models, including the rishon model. In addition to their electric charge and color the rishons are assumed to have hypercolor and the antirishons to have antihypercolor. Only combinations of three rishons or three antirishons are allowed because only those combinations are neutral with respect to hypercolor. A mixed three-particle system, such as $TT\overline{T}$, cannot exist because it would not be hypercolorless. The assignment of hypercolors thereby explains the rule for forming composite rishon systems. Similar rules apply in other hypercolor-based prequark models.

If the aim of a prequark model is to simplify the understanding of nature, postulating a new basic force does not seem very helpful. In the case of hypercolor, however, there may be some compensation. Consider the neutrino: it has neither electric charge nor color, only weak charge. According to the standard model, two neutrinos can act on each other only through the short-range weak force. If neutrinos are composites of hypercolored prequarks, however, there could be an additional source of interactions between neutrinos. When two neutrinos are far apart, there are practically no hypercolor forces between them, but when they are at close range, the hypercolored prequarks inside one

neutrino are able to "see" the inner hypercolors of the other one. Complicated short-range attractions and repulsions are the result. The mechanism, of course, is exactly the same as the one that explains the molecular force as a residue of the electromagnetic force and the strong force as a residue of the color force.

The conclusion may also be the same. Seiberg and I, and independently Greenberg and Sucher, were the first to suggest that the short-range weak force may actually be a residual effect of the hypercolor force. According to this hypothesis, the weak bosons W^+, W^- and Z^0 must also be composite objects, presumably made up of certain combinations of the same prequarks that compose the quarks and leptons. If this idea is confirmed, the list of fundamental forces will still have four entries: gravitation, electromagnetism, color and hypercolor. It should be noted, however, that all these forces are long-range ones; the short-range molecular, strong and weak forces will have lost their fundamental status.

For now hypercolor remains a conjecture, and so does the notion of explaining the weak force as a residue of the hypercolor force. It may yet turn out that the weak force is fundamental. A careful measurement of the mass, lifetime and other properties of the weak bosons should provide important clues in this matter.

Hypercolor is not the only candidate for a prequark binding force. Another interesting possibility was suggested by Pati, Salam and Strathdee. Instead of introducing a new hypercolor force, they borrowed an idea that has long been familiar, namely the magnetic force, and adapted it to a new purpose. An ordinary magnet invariably has two poles, which can be thought of as opposite magnetic charges. For 50 years there have been theoretical reasons for supposing there could also be isolated magnetic charges, or monopoles. Pati, Salam and Strathdee have argued that the prequarks could be particles with charges resembling both magnetic and electric charges. If they are, the forces binding them may be of a new and interesting origin.

None of the ideas I have just described constitutes a theory of prequark dynamics. Indeed, there is a serious impediment to the formulation of such a theory; it is the requirement that the prequarks be exceedingly small. The most stringent limit on their size is set indirectly by measurements of the magnetic moment of the electron, which agree with the calculations of quantum electrodynamics to an accuracy of 10 significant digits. In the calculations it is assumed that the electron is point-like; if it had any spatial extension or internal structure, the measured value would differ from the calculated one. Evidently any such discrepancy can at most affect the 11th digit of the result. It is this constraint that implies the characteristic distance scale of the electron's internal structure must be less than 10^{-16} centimeter. Roughly speaking, that is the maximum radius of an electron, and any prequarks must stay within it. If they strayed any wider, their presence would already have been detected.

Why should the small size of the electron inhibit speculation about its internal structure? The uncertainty principle establishes a reciprocal relation be-tween the size of a composite system and the kinetic energy of any components moving inside it. The smaller the system, the larger the kinetic energy of the constituents (see Figure 12.9). It follows that the prequarks must have enormous energy: more than 100 GeV (100 billion electron volts), and possibly much more. (One electron volt is the energy acquired by an electron when it is accelerated through a potential difference of one volt.) Because mass is fundamentally equivalent to energy, it can be measured in the same system of units. The mass of the electron, for example, is equivalent to .0005 GeV. There is a paradox here, which I call the energy mismatch: the mass of the composite system (if it is

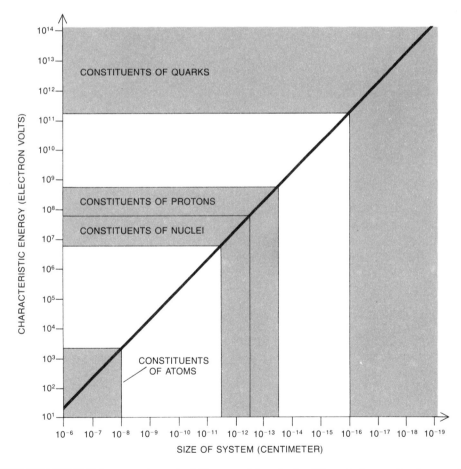

Figure 12.9 SIZE AND ENERGY have a reciprocal relation in the quantum theory, indicating that the constituents of composite quarks and leptons must have an exceedingly high kinetic energy. In a nucleus the protons and neutrons move with an energy of several million electron volts, and in a proton or a neutron the quarks have energies of several hundred million electron volts. Any constituents of quarks and leptons must be confined to a radius of less than 10^{-16} centimeter, and possibly much less. As a result the kinetic energy of the hypothetical prequarks can be no less than a few hundred billion electron volts.

indeed composite) is much smaller than the energy of its constituents.

The oddity of the situation can be illuminated by considering the relations of mass and kinetic energy in other composite systems. In an atom the kinetic energy of a typical electron is smaller than the mass of the atom by many orders of magnitude. In hydrogen, for example, the ratio is roughly one part in 100 million. The energy needed to change the orbit of the electron and thereby put the atom into an excited state is likewise a negligible fraction of the atomic mass. In a nucleus the kinetic energy of the protons and neutrons is also small compared with the nuclear mass, but it is not completely negligible. The motion of the particles gives them an energy equivalent to about 1 percent of the system's mass. The energy needed to create an excited state is also about 1 percent of the mass.

With the proton and its quark constituents the energy-mass relation begins to get curious. From the effective radius of the proton the typical energy of its component quarks can be calculated; it turns out to be comparable to the mass of the proton itself, which is a little less than 1 GeV. The energy that must be invested to create an excited state of the quark system is of the same order of magnitude: the hadrons identified as excited states of the proton exceed it in mass by from 30 to 100 percent. Nevertheless, the ratio of kinetic energy to total mass is still in the range that seems intuitively reasonable. Suppose one knew only the radius of the proton, and hence the typical energy of whatever happens to be inside it, and one were asked to guess the proton's mass. Since the energy of the constituents is generally a few hundred million electron volts, one would surely guess that the total mass of the system is at least of the same order of magnitude and possibly greater. The guess would be correct.

For the atom, the nucleus and the proton, then, the mass of the system is at least as large as the kinetic energy of the constituents and in some cases is much larger. If quarks and leptons are composite, however, the relation of energy to mass must be quite different. Since the prequarks have energies well above 100 GeV, one would guess that they would form composites with masses of hundreds of GeV or more. Actually the known quarks and leptons have masses that are much smaller; in the case of the electron and the neutrinos the mass is smaller by at least six orders of magnitude. The whole is much less than the sum of its parts.

The high energy of the prequarks is also what spoils the idea of viewing the higher generations of quarks and leptons as excited states of the same set of prequarks that form the first-generation particles. As in the other composite systems, the energy needed to change the orbits of the prequarks should be of the same order of magnitude as the kinetic energy of the constituents. One would therefore expect the successive generations to differ in mass by hundreds of GeV, whereas the actual mass differences are as small as .1 GeV.

At this point one might well adopt the view that the energy mismatch cannot be accepted, indeed that it simply demonstrates the elementary and structureless nature of the quarks and leptons. Many physicists hold this view. The energy mismatch, however, contradicts no basic law of physics, and I would argue that the circumstantial evidence for quark and lepton compositeness is sufficiently persuasive to warrant further investigation.

What is peculiar about the quark and lepton masses is not merely that they are small but that they are virtually zero when measured on the energy scale defined by their constituents' energy. In other composite systems a small amount of mass is "lost" by being converted into the binding energy of the system. The total mass of a hydrogen atom, for example, is slightly less than that of an isolated proton and electron; the difference is equal to the binding energy. In a nucleus this "mass defect" can reach a few percent of the total mass. In a quark or a lepton, it seems, the entire mass of the system is canceled almost exactly. Such a "miraculous" cancellation is certainly not impossible, but it seems most unlikely to happen by accident. Similar large cancellations are known elsewhere in physics, and they have always been found to result from some symmetry principle or conservation law. If there is to be any hope of constructing a theory of prequark dynamics, it is essential to find such a symmetry in this case.

There is a likely candidate: chiral symmetry, or chirality (see Figure 12.11). The name is derived from the Greek word for hand, and the symmetry has to do with handedness, the property defined by a particle's spin and direction of motion. Like other symmetries of nature, chiral symmetry has a conservation law associated with it, which gives the clearest account of what the symmetry means. The law states that the total number of right-handed

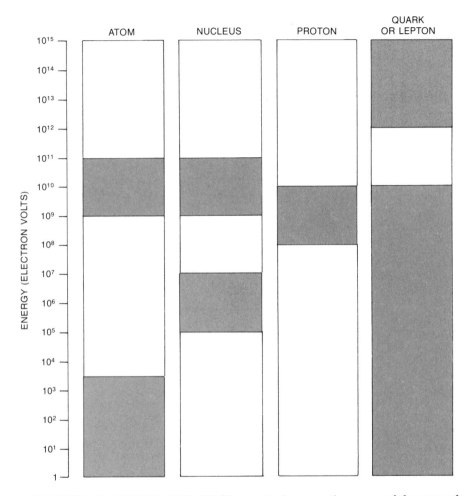

Figure 12.10 MISMATCH OF ENERGY AND MASS makes it difficult to devise a theory of how prequarks might move and interact. In an atom or a nucleus the kinetic energy of the constituents (*color*) is much less than the total mass of the system (*gray*). In the proton the two quantities are of comparable magnitude. In a composite quark, however, the energy of the prequarks greatly exceeds the total mass. Indeed, compared with the kinetic energy, the mass is essentially zero. Somehow virtually all mass is canceled, a development that is unlikely to be accidental.

particles and the total number of left-handed ones can never change.

In the ordinary world of protons, electrons and similar particles handedness or chirality clearly is not conserved. A violation of the conservation law can be demonstrated by a simple thought experiment. Imagine that an observer is moving in a straight line when he is overtaken by an electron. As the electron recedes from him he notes that its spin and direction of motion are related by a right-hand rule. Now suppose the observer speeds up, so that he is overtaking the electron. In the observer's frame of reference the electron seems to be approaching; in other words, it has reversed direction. Because its spin has not changed, however, it has become a left-handed particle.

There is one kind of particle to which this thought experiment cannot be applied: a massless particle. Because a massless particle must always move with the speed of light, no observer can ever go faster. As a result the handedness of a massless particle is an invariant property, independent of the observer's

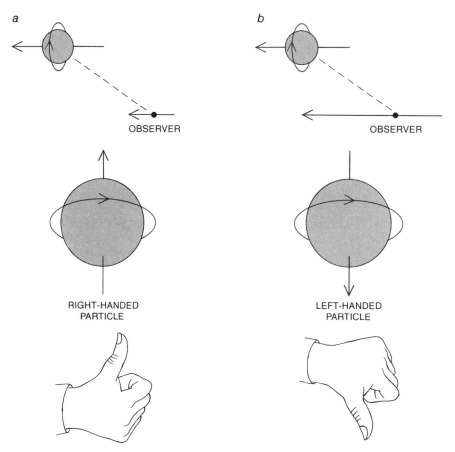

Figure 12.11 CHIRALITY, or handedness, describes the relation of particle's spin angular momentum to its direction of motion. If an observer is overtaken by a faster-moving electron (*a*), the electron obeys a right-hand rule: When the fingers of the right hand curl in the same direction as the spin, the thumb gives the direction of motion. If the observer overtakes the electron (*b*), the handedness of the particle changes because, in the observer's frame of reference, the electron is now approaching but its spin direction has not changed. If a theory of prequarks had a chiral symmetry, the low mass of the quarks and leptons might not be accidental. They would have to be virtually massless for the chiral symmetry to be maintained.

frame of reference. Furthermore, it can be shown that none of the known forces of nature (those mediated by the photon, the gluons and the weak bosons) can alter the handedness of a particle. Thus if the world were made up exclusively of massless particles, the world could be said to have chiral symmetry.

Chiral symmetry is the root of an idea that might conceivably account for the small mass of the quarks and leptons. The argument runs as follows. If the prequarks are massless particles, if they have a spin of 1/2 and if they interact with one another only through the exchange of gauge bosons, any theory describing their motion is guaranteed to have a chiral symmetry. If the massless prequarks then bind together to form composite spin-1/2 objects (namely the quarks and leptons), the chiral symmetry might ensure that the composite particles also remain massless compared with the huge energy of the prequarks inside them. Hence the small mass of the quarks and leptons is not an accident. They must be essentially massless with respect to the energy of their constituents if the chiral symmetry of the theory is to be maintained.

The crucial step in this argument is the one extending the chiral symmetry from a world of massless prequarks to one made up of composite quarks and leptons. It is essential that the symmetry of the original physical system survive in and be respected by the composite states formed out of the massless constituents. It may seem self-evident that if a theory is symmetrical in some sense, the physical systems described by the theory must exhibit that symmetry; actually, however, the spontaneous breaking of symmetries is commonplace. A familiar example is the roulette wheel. A physical theory of the roulette wheel would show it is completely symmetrical in the sense that each slot is equivalent to any other slot. The physical system formed by putting a ball in the roulette wheel, however, is decidedly asymmetrical: the ball invariably comes to rest in just one slot.

In the standard model it is the spontaneous breaking of a symmetry that makes the three weak bosons massive and leaves the photon massless. The theory that describes these gauge bosons is symmetrical, and in it the four bosons are essentially indistinguishable, but because of the symmetry breaking the physical states actually observed are quite different. Chiral symmetries are notoriously susceptible to symmetry breaking. Whether the chiral symmetry of prequarks breaks or not when the prequarks form composite objects can be determined only with a detailed understanding of the forces acting on the prequarks. For now that understanding does not exist. In certain models it can be shown that a chiral symmetry does exist but is definitely broken. No one has yet succeeded in constructing a composite model of quarks and leptons in which a chiral symmetry is known to remain unbroken. Neither the preon model nor the rishon model succeeds in solving the problem. The task is probably the most difficult one facing those attempting to demonstrate that quarks and leptons are composite.

If a consistent prequark theory can be worked out, it will still have to pass the test of experiment. First, it is important to establish in the laboratory whether or not quarks and leptons have any internal structure at all. If they do, experiments might then begin to discriminate among the various models. The experiments will have to penetrate the unknown realm of distances smaller than 10^{-16} centimeter and energies higher than 100 GeV. There are two basic ways to explore this region: by doing experiments with particles accelerated to very high energy and by making precise measurements of low-energy quantities that depend on the physics of events at very small distances.

Experiments of the first kind include the investigation of the weak bosons and the search for the Higgs particles of the standard model. When such particles can be made in sufficient numbers, a careful look at their properties should reveal much about the physics of very small distances. New ac-

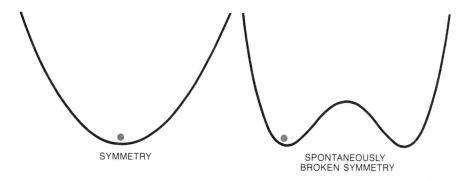

SYMMETRY

SPONTANEOUSLY
BROKEN SYMMETRY

Figure 12.12 SPONTANEOUS SYMMETRY BREAKING could spoil a prequark theory even if it has a chiral symmetry. Both physical systems shown here—a simple trough and a trough with a bump—are symmetrical in the sense that exchanging left and right leaves the system unaltered. For the simple trough the system remains symmetrical when a ball is put in the trough; the ball comes to rest in the center, so that exchanging left and right still has no effect. In the trough with a bump, the ball takes up a position on one side, and the symmetry is inevitably broken. Similarly, a prequark theory that has a chiral symmetry might nonetheless give rise to composite systems that do not observe the symmetry.

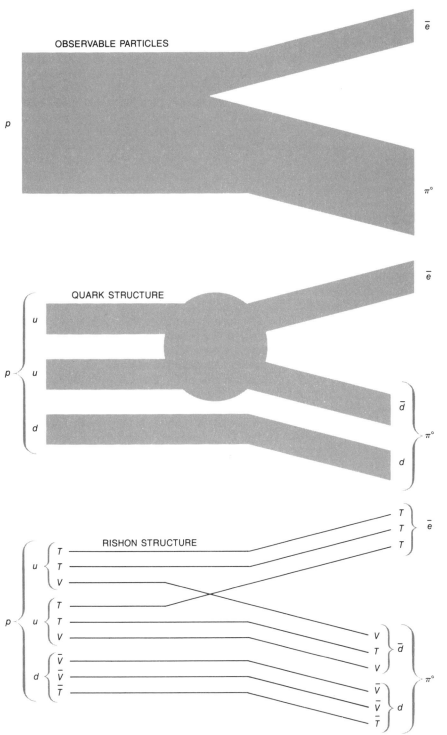

Figure 12.13 **CONJEC-TURED DECAY OF THE PROTON.** In one possible decay the proton would be observed to disintegrate into a positron ($\bar{e}$) and a neutral pion (π^0). The event can be understood in terms of the proton's quark constituents: an interaction of the two u quarks converts one of them into a positron and the other into a $\bar{d}$ antiquark; the latter combines with the remaining $\bar{d}$ quark of the proton to form the neutral pion. Grand unified theories suggest that the interaction of the u quarks is mediated by a new force of nature. The rishon model provides an alternative explanation: the two u quarks merely exchange a T and a V rishon.

celerators now being planned or built in the U.S., Europe and Japan are expected to yield detailed information about the weak bosons and will also continue the ongoing investigation of the quarks and leptons themselves.

Equally interesting are the high-precision, low-energy experiments. One of these is the search for the decay of the proton, a particle that is known to have an average lifetime of at least 10^{30} years (see Figure 12.13). Several experiments are now monitoring large quantities of matter, incorporating substantially more than 10^{30} protons, in an attempt to detect the signals emitted when a proton disintegrates. None of the forces of the standard model can induce such an event, but none of the rules of the standard model absolutely forbids it. Both the grand unified theories and the prequark models, on the other hand, include mechanisms that could convert a proton into other particles that would ultimately leave behind only leptons and photons. If the decay is detected, its rate and the pattern of decay products could offer an important glimpse beyond the standard model.

There is similar interest in the hypothetical process in which a muon emits a photon and is thereby converted into an electron. Again none of the forces of the standard model can bring about an event of this kind, but again too no fundamental law forbids it. Some of the composite models allow the transition and others do not, so that a search for the process might offer a means of choosing among the models. Experiments done up to now put a limit of less than one in 10 billion on the probability that any given muon will decay in this way. Detection of such events and a determination of their rate might illuminate the mysterious distinction between the generations.

A third class of precision experiments are those that continue to refine the measurement of the magnetic moment of the electron and of the muon. Further improvements can be expected both in experimental accuracy and in the associated calculations of quantum electrodynamics. If the results continue to agree with the predictions of the standard model, the limit on the possible size of any quark and lepton substructure will become remoter. If a discrepancy between theory and experiment is detected, it will represent a strong hint that quarks and leptons are not elementary.

It may well be a decade or two before the next level in the structure of matter comes clearly into view (if, again, there is another level). What is needed is a sound theoretical picture, one that is self-consistent, that agrees with all experiments and that is simple enough to explain all the features of the standard model in terms of a few principles and a few fundamental particles and forces. The correct picture, whether it is a grand unified theory or a composite model of the quarks and leptons, may already exist in some embryonic form. On the other hand, it is also possible the correct theory will emerge only from some totally new idea. In the words of Niels Bohr, it may be that our present ideas "are not sufficiently crazy to be correct."

Postscript: Section V,
Now and Beyond

. . .

Richard A. Carrigan, Jr., and W. Peter Trower

A "standard model" has evolved for the theory linking electromagnetism and radio-activity, and formulations exist for including the nuclear forces. While there remain mysteries such as the absence of the Higgs particle, the potential to find the Higgs with new accelerators such as the Superconducting Super Collider (SSC) is now within reach.

This beautiful picture is not without flaws. One of the most telling is that there are too many particles and too many forces. Nature, so it is argued, should not be so obtuse as to require four forces, more than twenty adjustable parameters and some sixty particles.

Are there ways that the picture could be simplified? Chapters 11 and 12 represent two of the most promising current efforts to look for physics beyond this standard picture.

Michael Green, along with John Schwarz, are the architects of the modern theory of superstrings. This theory, constructed in a ten-dimensional space, permits the inclusion of gravity. Curling up six of the dimensions in very tight loops eliminates most of the impact on macrophysics as we know it, while curing the problems with black holes that arise in "ordinary" quantum gravity. There nature is no longer dominated by point particles but by tiny strings 10^{-35} centimeters long, a billion-trillion times smaller than the atomic nucleus. The theory also accommodates the supersymmetry picture mentioned in Chapter 1 by Chris Quigg.

At first this appears to be a wide open theory. However, it turns out to be tightly constrained, as there are no adjustable parameters and a small menu of theoretical possibilities. A powerful arsenal of new and challenging mathematics is required to implement the theory. For the past several years there has been a veritable explosion of activity in this field as one development has followed another. One interesting possibility is the existence of shadow matter visible only through its gravitational interactions. The theory addresses some interesting questions like why is space-time four dimensional, and why does the cosmological constant vanish? Theoretical activity is intense, but there are no concrete experimental predictions so far.

The theory of subquarks presented in Chapter 12 by Haim Harari is much more familiar. We know atoms consist of more fundamental electrons and protons, and protons consist of still more fundamental quarks. So what are quarks made of? The obvious answer is subquarks — big fleas must have little fleas. Harari espouses such pictures with sub-quarks or preons.

The models, however, have problems. The new subparticles must be bound inside an electron or a quark by some new fundamental force. They must move inside the electron or the quark with very large energies — at least hundreds of GeV's — while the total mass of the system is much smaller. This can happen only if the new force obeys some new principles, presumably involving new symmetries of nature. In the absence of direct experimental evidence for substructure, it is extremely difficult to learn anything about the possible new force and new symmetries. As a result, there have not been many developments since 1983.

Green and Harari point to two different directions the physics of the future might take. Other ideas will come and theories that seem different may later merge. It is the process of modern physics, where unimagined linkages appear, new questions are asked, more facts arise and suddenly there is an exciting leap forward. This physics is a wonderful amalgam of theory, accelerator science, cosmology, mathematics and experiment. We stand at the edge of an ocean and confront an unknown sea stretching to the horizon. To our backs lie the richly explored terrain of our efforts. At the shore a vigorous swimmer plunges in and in an instant knows more about that mysterious ocean than anyone who has gone before.

The Authors

THE EDITORS

RICHARD A. CARRIGAN, JR., and W. PETER TROWER are experimental particle physicists. They both received their Ph.D.'s from the University of Illinois. Carrigan is head of Fermilab's Office of Research and Technology Applications. Trower is on the faculty of the Virginia Polytechnic Institute and State University.

CHRIS QUIGG ("Elementary Particles and Forces") is deputy head of the Superconducting Super Collider Central Design Group at the Lawrence Berkeley Laboratory, Berkeley, California. He is a graduate of Yale University and the University of California at Berkeley.

SHELDON LEE GLASHOW ("Quarks with Color and Flavor") is at Harvard University. An alumnus of the Bronx High School of Science, he received his Ph.D. from Harvard. He shared the Nobel Prize in physics in 1979.

CLAUDIO REBBI ("The Lattice Theory of Quark Confinement") is a theorist on the staff of the Brookhaven National Laboratory. Born in Trieste, he received his degree from the University of Torino.

ROBERT R. WILSON ("The Next Generation of Particle Accelerators"), a leading figure in high-energy particle accelerators and founding director of Fermilab, obtained his degrees at the University of California at Berkeley and is a Professor Emeritus at Columbia, Cornell, and Chicago Universities.

J. DAVID JACKSON, MAURY TIGNER and STANLEY WOJCICKI ("The Superconducting Super Collider") have been major figures in the development of the SSC design. Jackson, a theoretical physicist, is on the faculty of the University of California at Berkeley. He has a Ph.D. from the Massachusetts Institute of Technology. Tigner, an accelerator designer, is on the faculty of Cornell University, where he obtained his Ph.D. Wojcicki, an experimental physicist, is on the physics faculty at Stanford University. He received his Ph.D. from the University of California at Berkeley.

MARTIN L. PERL and WILLIAM T. KIRK ("Heavy Leptons") are at the Stanford Linear Accelerator Center. Perl obtained his Ph.D. in physics from Columbia University. He was team leader in the discovery of the tau lepton and shared the 1982 Wolf Prize in physics. Kirk, a public relations officer, received his bachelor's degree in history at Cornell University.

DAVID B. CLINE, CARLO RUBBIA and SIMON VAN DER MEER ("The Search for Intermediate Vector Bosons"). Cline, who is on the faculty at the University of California at Los Angeles, has a Ph.D. from the University of Wisconsin. Rubbia was trained at the Normal School of Pisa and the University of Pisa, Italy. He is Director General of CERN. Van der Meer, an accelerator builder, received his diploma in engineering from the University of Guelph and is on the staff at CERN. Rubbia and van der Meer shared the 1984 Nobel Prize in physics for their discovery of the intermediate bosons.

LEON M. LEDERMAN ("The Upsilon Particle"), director of Fermilab, did his undergraduate work at the City College of New York and received his Ph.D. from Columbia University. For his discovery of a second type of neutrino, he shared the 1988 Nobel Prize in physics, and for his work in discovering the upsilon particle, he shared the 1982 Wolf Prize with Martin Perl.

ELLIOT D. BLOOM and GARY J. FELDMAN ("Quarkonium") are on the faculty at the Stanford Linear Accelerator Center. Bloom's Ph.D. is from the California Institute of Technology. Feldman received his doctorate from Harvard University.

NARIMAN B. MISTRY, RONALD A. POLING and EDWARD H. THORNDIKE ("Particles with Naked Beauty") are members of the CLEO collaboration at

Cornell. Mistry, now on the staff of Cornell University, received his Ph.D. from Columbia University. Poling is on the staff at the University of Rochester, where he received his Ph.D. Thorndike, on the physics faculty at the University of Rochester, earned his Ph.D. at Harvard University.

MICHAEL B. GREEN ("Superstrings") is professor of physics at Queen Mary College of the University of London and obtained his Ph.D. from the University of Cambridge.

HAIM HARARI ("The Structure of Quarks and Leptons"), a professor at the Weizmann Institute in Israel, received his degrees from the Hebrew University in Jerusalem.

Bibliography

Comprehensive readings:

Cahn, Robert N., and Gerson Goldhaber. 1988. *The experimental foundations of particle physics*. Cambridge: Cambridge University Press.

Carrigan, Richard A., Jr., and W. Peter Trower. 1989. *Particle physics in the cosmos*. New York: W. H. Freeman and Company.

Close, Frank. 1986. *The cosmic onion*. New York: American Institute of Physics.

Close, Frank, Michael Martin and Christine Sutton. 1987. *The particle explosion*. Oxford: Oxford University Press.

Cooper, Mecia Grant, and Geoggrey B. West. 1987. *Particle physics: A Los Alamos primer*. Cambridge: Cambridge University Press.

Dodd, James. 1984. *The ideas of particle physics: An introduction for scientists*. Cambridge: Cambridge University Press.

Duff, Brian G. 1986. *Fundamental particles: An introduction to quarks and leptons*. London: Taylor and Francis.

Grease, Robert P., and Charles C. Mann. 1986. *The second creation*. New York: Macmillan.

Ne'eman, Yuval, and Yoram Kirsh. 1986. *The particle hunters*. Cambridge: Cambridge University Press.

Pais, Abraham. 1986. *Inward bound: Of matter and forces in the physical world*. Oxford: Oxford University Press.

Riordan, Michael. 1987. *The hunting of the quark*. New York: Simon and Schuster.

Taubes, Gary. 1986. *Nobel dreams: Power, deceit and the ultimate experiment*. New York: Random House.

Weinberg, Steven. 1984. *The discovery of subatomic particles*. New York: W. H. Freeman and Company.

Wilczek, Frank, and Betsy Devine. 1988. *Longing for the harmonies: Themes and variations from modern physics*. New York: Norton.

Zee, Anthony. 1986. *Fearful symmetry: The search for beauty in modern physics*. New York: Macmillan.

Readings from the magazine:

Adair, Robert K. 1988. A flaw in the universal mirror. *Scientific American* 258 (February): 50–56.

Carrigan, Richard A., JR., and W. Peter Trower. 1982. Superheavy magnetic monopoles. *Scientific American* 246 (April): 106–118.

Cline, D. B. 1988. Beyond truth and beauty: A fourth family of particles. *Scientific American* 259 (August): 60–71.

Freedman, Daniel Z., and Peter Van Nieuwenhuizen. 1985. The hidden dimensions of space-time. *Scientific American* 252 (March): 74–81.

Georgi, Howard. 1981. A unified theory of elementary particles and forces. *Scientific American* 244 (April): 48–63.

Haber, Howard E., and Gordon L. Kane. 1986. Is nature supersymmetric? *Scientific American* 254 (June): 52–60.

't Hooft, Gerard. 1980. Gauge theories of the forces between elementary particles. *Scientific American* 242 (June): 104–138.

LoSecco, J. M., Frederick Reines and Daniel Sinclair. 1985. The search for proton decay. *Scientific American* 252 (June): 54–62.

Schramm, D. M., and G. Steigman. 1988. Particle accelerators test cosmological theory. *Scientific American* 258 (June): 66–72.

Veltman, Martinus J. G. 1986. The Higgs boson. *Scientific American* 255 (November): 76–84.

Weinberg, Steven. 1981. The decay of the proton. *Scientific American* 244 (June): 64–75.

Original literature for the advanced reader:

Annual review of astronomy and astrophysics. Palo Alto, Calif.: Annual Reviews.

Annual review of nuclear and particle science. Palo Alto, Calif.: Annual Reviews.

INDEX

Page numbers in *italics* indicate illustrations.